The Ultimate Doily Book

Bobbie Matela, Managing Editor

Carol Wilson Mansfield, Art Director

Mary Ann Frits, Editorial Director

Kelly Robinson, Sandy Scoville, and Kathy Wesley, Pattern Editors

Graphic Solutions inc-chgo, Book Design

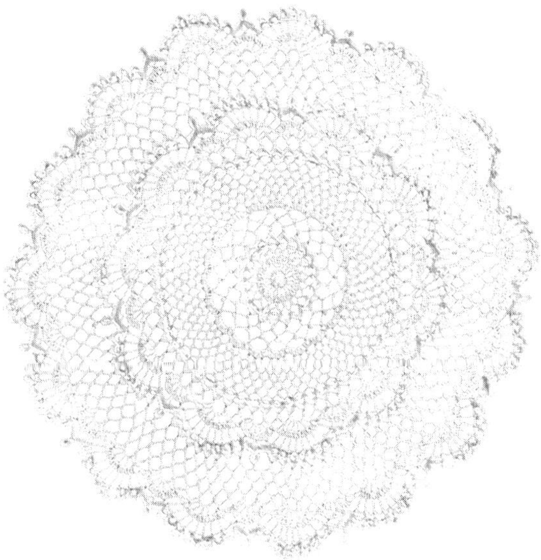

Patterns tested and models made by Margaret Gordon, Terry Gunter, Marlene Hommerding, Joey Koening, Lori Llewellyn, Dolores Roberts, Kelly Robinson, and Sandy Scoville.

ISBN: 0-88195-597-3

Printed in U.S.A.

7 8 9

Introduction

This wonderful collection of seventeen doilies is presented as a tribute to the designers of the past from the designers of today.

Doilies bridge the centuries, from the 1800's through the 1900's and are still going strong as we approach the 2000's.

Not all doily patterns are old; in fact, this entire collection was created just for this book by four contemporary designers, though their classic designs and timeless beauty are firmly rooted in the past.

Doilies were a practical and fashionable accent in Victorian homes. They were named after a famous French draper, Mr. D'Oyley, who encouraged their use to protect fine upholstery from the greasy hair oil sported by Victorian dandies.

Mr. D'Oyley has left a legacy of lovely, lace-like creations that are as warm and welcoming in the modern home as they were over a hundred years ago. And we owe our thanks to Lucille LaFlamme, Jo Ann Maxwell, Linda Mershon, Kathy Wigington, and Yalanda Wiese for carrying on the tradition of beautiful design.

General Directions

There are several basic techniques and information on materials that will help you in creating our doilies. We have included this information here. We suggest that you read through these General Directions carefully before beginning your doily.

Crochet Materials

Steel hooks: If you have never worked with steel hooks and finer thread, don't be afraid to try. You will be using exactly the same stitches you're familiar with, but at first it will feel clumsy and awkward. For an experienced crocheter, this is a bit of a surprise -- suddenly feeling all thumbs again just as when you first learned to crochet. But this will pass in a few hours of crocheting, as you adjust your tension and working method to the new tools. Soon you will work much more by feel than when working with heavier yarns. So be patient with any initial clumsiness and confusion -- they won't be with you long.

Steel hooks range in size from 00 (large) to 14 (very fine), and are 5" long which is shorter than the aluminum or plastic hooks. Their shape is different from the other crochet hooks. There is the throat, then the shank, and after the shank the steel begins to widen again before it reaches the finger grip (**Fig 1**). When crocheting, it is important that the stitches do not slide beyond the shank as this will cause a loose tension and alter the gauge. If you find you are having difficulty at first, put a piece of cellophane tape around the hook to keep the stitches from sliding past the correct area. With practice, you will work in the right place automatically.

Fig 1

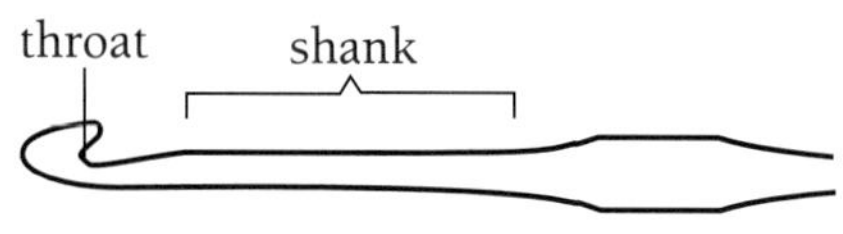

Crochet cotton: Bedspread-weight crochet cotton, the size usually used to make tablecloths and bedspreads, is used for the doilies in this book. It is sold under various brand names, such as: J. & P. Coats "Knit-Cro-Sheen®", "South Maid", or "Opera" size 10; Anchor "Crochet Cotton", size 10; DMC "Cebelia" size 10; or Grandma's Best.

Size and Gauge

It is important to achieve the gauge specified, otherwise your doily will not be the correct size. Please take the time to work a stitch gauge swatch, about 4" x 4". Measure the swatch. If the number of stitches are fewer than indicated under "Gauge" in the pattern, your hook is too large. Try another swatch with a smaller size hook. If the number of stitches are more than indicated under "Gauge" in the pattern, your hook is too small. Try another swatch with a larger size hook.

Laundering

Wash doily carefully by hand using a mild soap. Rinse well in warm water. Spread it out to dry on a flat padded surface following instructions below.

Blocking

This means "setting" the finished doily into its final size and shape. To do this, spread doily out on a flat padded surface, having right side facing up. Dampen with water (if necessary). Smooth project out to correct size, having the design properly aligned with all picots, loops, etc. open. If necessary, use rust-proof pins to hold the edges in place. Let dry thoroughly before removing.

Abbreviations and Symbols

beg	begin(ning)
bl(s)	back loop(s)
ch(s)	chain(s)
CL(s)	cluster(s)
dc	double crochet(s)
dec	decrease(-ing)
dtrc	double triple crochet(s)
hdc	half double crochet(s)
LK(s)	lover's knot(s)
lp(s)	loop(s)
patt	pattern
prev	previous
rem	remain(ing)
rep	repeat(ing)
rnd(s)	round(s)
sc	single crochet(s)
sk	skip
sl	slip
sl st(s)	slip stitch(es)
sp(s)	space(s)
st(s)	stitch(es)
tog	together
tr trc	triple triple crochet(s)
trc	triple crochet(s)
YO	yarn over

* An asterisk is used to mark the beginning of a portion of instructions which will be worked more than once; thus, "rep from * twice" means after working the instructions once, repeat the instructions following the asterisk twice more (3 times in all).

† The dagger (or double daggers ††) identifies a portion of instructions that will be repeated again later in the same row or round.

: The number after a colon at the end of a row or round indicates the number of stitches you should have when the row or round has been completed.

() Parentheses are used to enclose instructions which should be worked the exact number of times specified immediately following the parentheses, such as (2 sc in next dc, sc in next dc) twice. They are also used to set off and clarify a group of stitches that are to be worked all into the same space or stitch, such as (2 dc, ch 1, 2 dc) in corner sp.

[] Brackets and () parentheses are used to provide additional information to clarify instructions.

Join - join with a sl st unless otherwise specified.

Front loop is the loop toward you at the top of the stitch.

Back loop is the loop away from you at the top to the stitch.

Post is the part of the stitch around which another stitch may be worked.

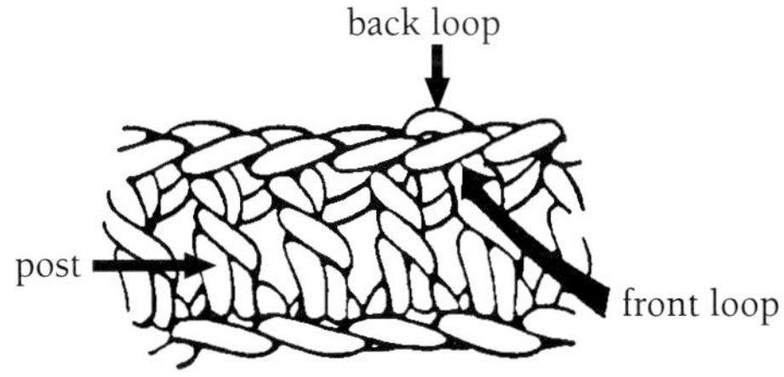

The patterns in this book are written using United States terminology. Terms which have different English equivalents are noted below.

United States	English
single crochet(sc)	double crochet(dc)
half double crochet(hdc)	half treble(htr)
double crochet(dc)	treble(tr)
triple crochet(trc)	double treble(dtr)
double triple crochet(dtrc)	triple treble(trtr)
triple triple crochet(tr trc)	quadruple treble(q[uad] trc)
skip(sk)	miss
slip stitch(sl st)	slip stitch(ss) or ("single crochet")
gauge	tension
yarn over(YO)	yarn over hook(YOH)
bedspread-weight crochet cotton	lightweight crochet cotton

Diamonds Are Forever

designed by Yalanda Wiese

Size:

About 17" diameter

Materials:

Bedspread-weight crochet cotton, two 225-yd balls white
Size 6 steel crochet hook, or size required for gauge

Gauge:

8 dc = 1"

Pattern Stitches

Cluster (CL)**:** Keeping last lp of each dc on hook, dc in next 2 sts, YO and draw through all 3 lps on hook: CL made.

Beg 3-double crochet Cluster (beg 3-dc CL)**:** Ch 2, keeping last lp of each dc on hook, dc in next dc and in next CL, YO and draw through all 3 lps on hook: beg 3-dc CL made.

3-double crochet Cluster (3-dc CL)**:** Keeping last lp of each dc on hook, dc in next 3 sts, YO and draw through all 3 lps on hook: 3-dc CL made.

Instructions

Ch 5, join to form a ring.

Rnd 1: Ch 5 (counts as a dc and a ch-2 sp on this and following rnds), * dc in ring, ch 2; rep from * 6 times more; join in 3rd ch of beg ch-5: 8 ch-2 sps.

Rnd 2: Sl st in next ch-2 sp; ch 3 (counts as a dc on this and following rnds), 2 dc in same sp; ch 2, * 3 dc in next ch-2 sp; ch 2; rep from * 6 times more; join in 3rd ch of beg ch-3.

Rnd 3: Ch 3, dc in same ch as joining and in next dc; 2 dc in next dc; ch 2, * 2 dc in next dc; dc in next dc, 2 dc in next dc; ch 2; rep from * 6 times more; join in 3rd ch of beg ch-3.

Rnd 4: Ch 3, dc in same ch as joining and in next 3 dc; 2 dc in next dc; ch 2, * 2 dc in next dc; dc in next 3 dc, 2 dc in next dc; ch 2; rep from * 6 times more; join in 3rd ch of beg ch-3.

Rnd 5: Ch 3, dc in same ch as joining and in next 5 dc; 2 dc in next dc; ch 2, * 2 dc in next dc; dc in next 5 dc, 2 dc in next dc; ch 2; rep from * 6 times more; join in 3rd ch of beg ch-3.

Rnd 6: Ch 3, dc in same ch as joining and in next 7 dc; 2 dc in next dc; ch 2, * 2 dc in next dc; dc in next 7 dc, 2 dc in next dc; ch 2; rep from * 6 times more; join in 3rd ch of beg ch-3.

Rnd 7: Ch 3, dc in same ch as joining; dc in next 9 dc, 2 dc in next dc; ch 2, * 2 dc in next dc; dc in next 9 dc, 2 dc in next dc; ch 2; rep from * 6 times more; join in 3rd ch of beg ch-3.

Rnd 8: Ch 2, dc in next dc: beg CL made; dc in next 9 dc, CL (see Pattern Stitches) over next 2 dc; ch 5, * CL over next 2 dc; dc in next 9 dc, CL over next 2 dc; ch 5; rep from * 6 times more; join in first dc.

Rnd 9: Beg CL; dc in next 7 dc, CL over next dc and next CL; ch 2, 5 dc in 3rd ch of next ch-5 lp; ch 2, * CL over next CL and next dc; dc in next 7 dc, CL over next dc and next CL; ch 2, 5 dc in 3rd ch of next ch-5 lp; ch 2; rep from * 6 times more; join in first dc.

Rnd 10: Beg CL; dc in next 5 dc, CL over next dc and next CL; ch 2, dc in next 2 dc, in next dc work (dc, ch 3, dc); dc in next 2 dc, ch 2, * CL over next CL and next dc; dc in next 5 dc, CL over next dc and next CL; ch 2, dc in next 2 dc, in next dc work (dc, ch 3, dc); dc in next 2 dc, ch 2; rep from * 6 times more; join in first dc.

Rnd 11: Beg CL; dc in next 3 dc, CL over next dc and next CL; ch 2, dc in next 3 dc, ch 3, dc in next ch-3 lp, ch 3, dc in next 3 dc, ch 2, * CL over next CL and next dc; dc in next 3 dc, CL over next dc and next CL; ch 2, dc in next 3 dc, ch 3, dc in next ch-3 lp, ch 3, dc in next 3 dc, ch 2; rep from * 6 times more; join in first dc.

Rnd 12: Beg CL; dc in next dc, CL over next dc and next CL; ch 2, dc in next 3 dc, ch 3, dc in next ch-3 lp, ch 2, dc in next ch-3 lp, ch 3, dc in next 3 dc, ch 2, * CL over next CL and next dc; dc in next dc, CL over next dc and next CL; ch 2, dc in next 3 dc, ch 3, dc in next ch-3 lp, ch 2, dc in next ch-3 lp, ch 3, dc in next 3 dc, ch 2; rep from * 6 times more; join in first dc.

Rnd 13: Beg 3-dc CL (see Pattern Stitches); ch 2, dc in next 3 dc, ch 3, dc in next ch-3 lp, ch 2, dc in next ch-2 sp, ch 2, dc in next ch-3 lp, ch 3, dc in next 3 dc, ch 2, * 3-dc CL (see Pattern Stitches) over next CL, next dc, and next CL; ch 2, dc in next 3 dc, ch 3, dc in next ch-3 lp, ch 2, dc in next ch-2 sp, ch 2, dc in next ch-3 lp, ch 3, dc in next 3 dc, ch 2; rep from * 6 times more; join in top of beg 3-dc CL.

Rnd 14: Ch 5, * dc in next 3 dc, ch 3, dc in next ch-3 lp, ch 2, (dc in next ch-2 sp, ch 2) twice; dc in next ch-3 lp, ch 3, dc in next 3 dc, ch 2, dc in top of next CL, ch 2; rep from * 7 times more, ending last rep without working last dc and ch-2 sp; join in 3rd ch of beg ch-5.

continued on page 8

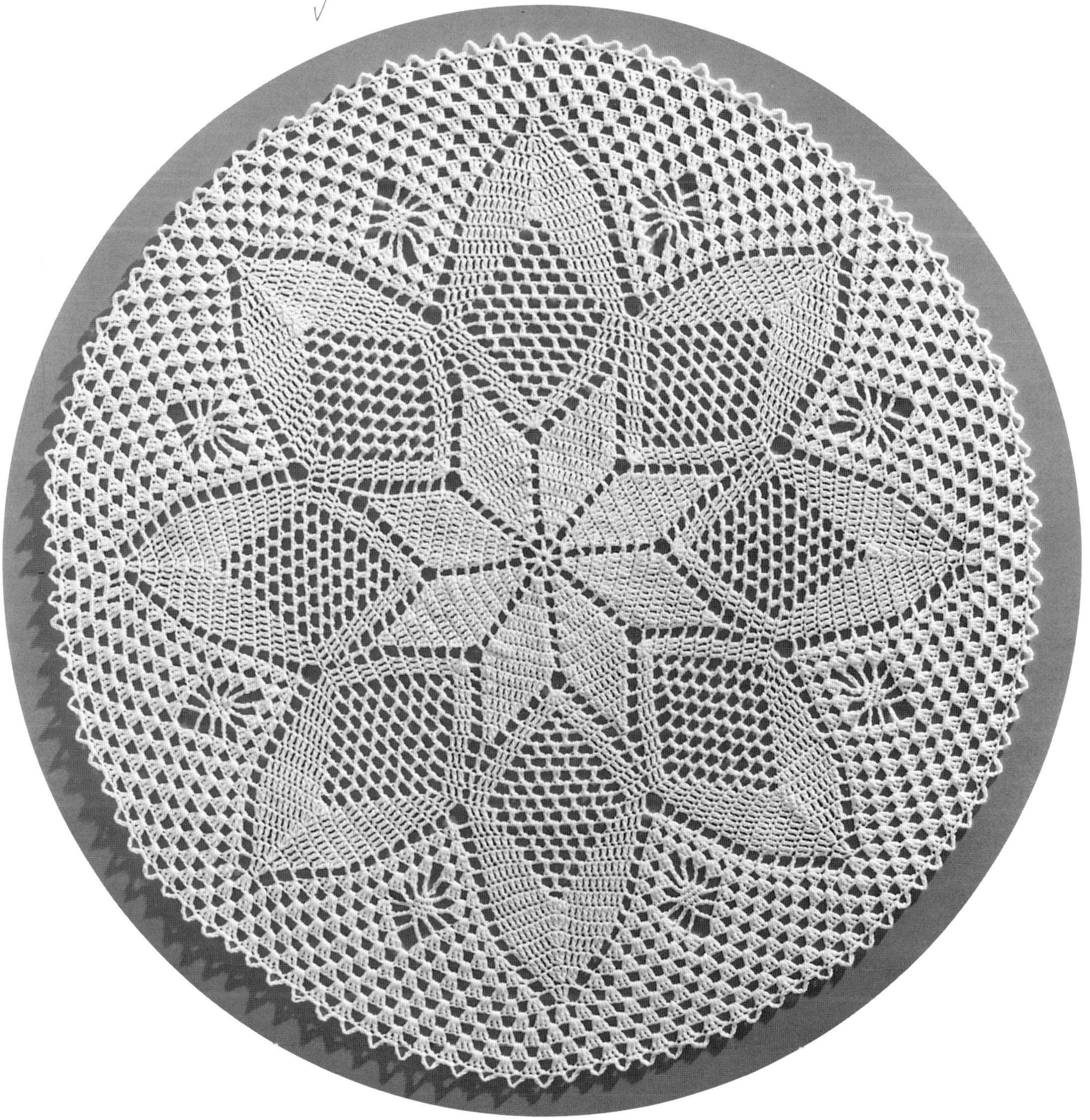

Rnd 15: Ch 4 (counts as a dc and a ch-1 sp), * dc in next 3 dc, ch 3, dc in next ch-3 lp, ch 2, (dc in next ch-2 sp, ch 2) 3 times; dc in next ch-3 lp, ch 3, dc in next 3 dc, ch 1, dc in next dc, ch 1; rep from * 7 times more, ending last rep without working last dc and ch 1; join in 3rd ch of beg ch-4.

Rnd 16: Ch 3, * dc in next 3 dc, ch 3, dc in next ch-3 lp, ch 2, (dc in next ch-2 sp, ch 2) 4 times; dc in next ch-3 lp, ch 3, dc in next 3 dc, dc in next dc; rep from * 7 times more, ending last rep without working last dc; join in 3rd ch of beg ch-3.

Rnd 17: Sl st in next dc; ch 3, dc in next 2 dc, * ch 3, dc in next ch-3 lp, ch 2, (dc in next ch-2 sp, ch 2) 5 times; dc in next ch-3 lp, ch 3, dc in next 3 dc, sk next dc, dc in next 3 dc; rep from * 7 times more, ending last rep without working last 3 dc; join in 3rd ch of beg ch-3.

Rnd 18: Ch 3, dc in next 2 dc, * dc in next ch-3 lp, ch 2, (dc in next ch-2 sp, ch 2) 6 times; dc in next ch-3 lp, dc in next 3 dc, ch 5, dc in next 3 dc; rep from * 7 times more, ending last rep without working last 3 dc; join in 3rd ch of beg ch-3.

Rnd 19: Ch 3, dc in next 3 dc, * (dc in next ch-2 sp, ch 2) 6 times; dc in next ch-2 sp, dc in next 4 dc, ch 2, 3 dc in 3rd ch of next ch-5 lp; ch 2, dc in next 4 dc; rep from * 7 times more, ending last rep without working last 4 dc; join in 3rd ch of beg ch-3.

Rnd 20: Ch 3, dc in next 4 dc, * (dc in next ch-2 sp, ch 2) 5 times; dc in next ch-2 sp, dc in next 5 dc, ch 2, 3 dc in next dc; ch 1, sk next dc, 3 dc in next dc; ch 2, dc in next 5 dc; rep from * 7 times more, ending last rep without working last 5 dc; join in 3rd ch of beg ch-3.

Rnd 21: Ch 3, dc in next 5 dc, * (dc in next ch-2 sp, ch 2) 4 times; dc in next ch-2 sp, dc in next 6 dc, ch 2, 3 dc in next dc; ch 1, 3 dc in next ch-1 sp; ch 1, sk next 2 dc, 3 dc in next dc; ch 2, dc in next 6 dc; rep from * 7 times more, ending last rep without working last 6 dc; join in 3rd ch of beg ch-3.

Rnd 22: Ch 3, dc in next 6 dc, * (dc in next ch-2 sp, ch 2) 3 times; dc in next ch-2 sp, dc in next 7 dc, ch 2, 3 dc in next dc; ch 1, 3 dc in next ch-1 sp; ch 7, 3 dc in next ch-1 sp; ch 1, sk next 2 dc, 3 dc in next dc; ch 2, dc in next 7 dc; rep from * 7 times more, ending last rep without working last 7 dc; join in 3rd ch of beg ch-3.

Rnd 23: Ch 3, dc in next 7 dc, * (dc in next ch-2 sp, ch 2) twice; dc in next ch-2 sp, dc in next 8 dc, ch 2, 3 dc in next dc; ch 1, 3 dc in next ch-1 sp; ch 3, sc in next ch-7 lp, ch 3, 3 dc in next ch-1 sp; ch 1, sk next 2 dc, 3 dc in next dc; ch 2, dc in next 8 dc; rep from * 7 times more, ending last rep without working last 8 dc; join in 3rd ch of beg ch-3.

Rnd 24: Ch 3, dc in next 8 dc, * dc in next ch-2 sp, ch 2, dc in next ch-2 sp, dc in next 9 dc, ch 2, 3 dc in next dc; ch 1, 3 dc in next ch-1 sp; ch 3, sc in next ch-3 lp close to sc of prev rnd, sc in next sc, sc in ch-3 lp, ch 3, 3 dc in next ch-1 sp; ch 1, sk next 2 dc, 3 dc in next dc; ch 2, dc in next 9 dc; rep from * 7 times more, ending last rep without working last 9 dc; join in 3rd ch of beg ch-3.

Rnd 25: Ch 3, dc in next 8 dc, * 3-dc CL over next dc, next ch-2 sp, and next dc; dc in next 9 dc, ch 2, 3 dc in next dc; ch 1, 3 dc in next ch-1 sp; ch 3, sc in next ch-3 lp close to sc of prev rnd, sc in next 3 sc, sc in next ch-3 lp, ch 3, 3 dc in next ch-1 sp; ch 1, sk next 2 dc, 3 dc in next dc; ch 2, dc in next 9 dc; rep from * 7 times more, ending last rep without working last 9 dc; join in 3rd ch of beg ch-3.

Rnd 26: Ch 2, dc in next 7 dc, * 3-dc CL over next dc, next 3-dc CL, and next dc; dc in next 6 dc, CL over next 2 dc; ch 2, 3 dc in next dc; ch 1, 3 dc in next ch-1 sp; ch 1, sk next 2 dc, 3 dc in next dc; ch 3, sk next sc, sc in next 3 sc, ch 3, 3 dc in next dc; ch 1, 3 dc in next ch-1 sp; ch 1, sk next 2 dc, 3 dc in next dc; ch 2, CL over next 2 dc; dc in next 6 dc; rep from * 7 times more, ending last rep without working last CL and last 6 dc; join in first dc.

Rnd 27: Ch 2, dc in next 5 dc, * 3-dc CL over next dc, next 3-dc CL, and next dc; dc in next 4 dc, CL over next dc and next CL; ch 2, 3 dc in next dc; ch 1, (3 dc in next ch-1 sp, ch 1) twice; sk next 2 dc, 3 dc in next dc; ch 3, sk next sc, sc in next sc, ch 3, 3 dc in next dc; ch 1, (3 dc in next ch-1 sp, ch 1) twice; sk next 2 dc, 3 dc in next dc; ch 2, CL over next CL and next dc; dc in next 4 dc; rep from * 7 times more, ending last rep without working last CL and last 4 dc; join in first dc.

Rnd 28: Ch 2, dc in next 3 dc, * 3-dc CL over next dc, next 3-dc CL, and next dc; dc in next 2 dc, CL over next dc and next CL; ch 2, 3 dc in next dc; ch 1, (3 dc in next ch-1 sp, ch 1) 3 times; sk next 2 dc, 3 dc in next dc; ch 1, 3 dc in next dc; ch 1, (3 dc in next ch-1 sp, ch 1) 3 times; sk next 2 dc, 3 dc in next dc; ch 2, CL over next CL and next dc; dc in next 2 dc; rep from * 7 times more, ending last rep without working last CL and last 2 dc; join in first dc.

Rnd 29: Ch 2, dc in next dc, * 3-dc CL over next dc, next 3-dc CL, and next dc; CL over next dc and next CL; ch 2, 3 dc in next dc; ch 1, (3 dc in next ch-1 sp, ch 1) 9 times; sk next 2 dc, 3 dc in next dc; ch 2, CL over next CL and next dc; rep from * 7 times more, ending last rep without working last CL; join in first dc.

Rnd 30: Ch 2, CL over next 3-dc CL and next CL; * ch 2, 3 dc in next dc; ch 1, (3 dc in next ch-1 sp, ch l) 10 times; sk next 2 dc, 3 dc in next dc; ch 2, 3-dc CL over next CL, next 3-dc CL, and next CL; rep from * 7 times more, ending last rep without working last 3-dc CL; join in top of first CL.

Rnd 31: Sl st in next 2 chs and in next dc; ch 3, 2 dc in same dc; * ch 1, (3 dc in next ch-1 sp, ch 1) 11 times; sk next 2 dc, 3 dc in next dc; ch 1, 3 dc in next CL; ch 1, 3 dc in next dc; rep from * 7 times more, ending last rep without working last 3 dc; join in 3rd ch of beg ch-3.

Rnd 32: Sl st in next 2 dc and in next ch-1 sp; ch 3, 2 dc in same ch-1 sp; ch 1, * 3 dc in next ch-1 sp, ch 1; rep from * around; join in 3rd ch of beg ch-3.

Rnds 33 and 34: Rep Rnd 32.

Rnd 35: Sl st in next 2 dc; ch 4, * sl st in next 3 dc; ch 4; rep from * around; join in joining sl st of prev rnd.

Finish off and weave in all ends.

Pineapple Sundae

designed by Yalanda Wiese

Size:

About 21" diameter

Materials:

Bedspread-weight crochet cotton, two 225-yd balls white
Size 6 steel crochet hook, or size required for gauge

Gauge:

8 dc = 1"

Pattern Stitches

Triple triple crochet (tr trc)**:** YO 4 times, draw up lp in next dc, (YO and draw through 2 lps on hook) 5 times: tr trc made.

Cluster (CL)**:** Keeping last lp of each trc on hook, trc in next 2 trc, YO and draw through all 3 lps on hook: CL made.

5-tr trc Cluster (5-tr trc CL)**:** Keeping last lp of each tr trc on hook, tr trc in next 5 trc, YO and draw through 6 lps on hook: 5-tr trc CL made.

8-trc Cluster (8-trc CL)**:** Keeping last lp of each trc on hook, trc in next 8 sts, YO and draw through all 9 lps on hook: 8-trc CL made.

Instructions

Ch 5, join to form a ring.

Rnd 1: Ch 1, (sc in ring, ch 10) 8 times; join in first sc: 8 ch-10 lps.

Rnd 2: Sl st in next 4 chs of next ch-10 lp and in same lp; ch 4 (counts as a trc on this and following rnds), 2 trc in same lp; ch 4, (3 trc in next ch-10 lp, ch 4) 7 times; join in 4th ch of beg ch-4.

Rnd 3: Ch 4, trc in same ch as joining and in next trc; 2 trc in next trc; ch 5, * 2 trc in next trc; trc in next trc, 2 trc in next trc; ch 5; rep from * 6 times more; join in 4th ch of beg ch-4.

Rnd 4: Ch 4, trc in same ch as joining and in next 3 trc; 2 trc in next trc; ch 5, * 2 trc in next trc; trc in next 3 trc, 2 trc in next trc; ch 5; rep from * 6 times more; join in 4th ch of beg ch-4.

Rnd 5: Ch 4, trc in same ch as joining and in next 5 trc; 2 trc in next trc; ch 5, * 2 trc in next trc; trc in next 5 trc, 2 trc in next trc; ch 5; rep from * 6 times more; join in 4th ch of beg ch-4.

Rnd 6: Ch 4, trc in same ch as joining and in next 7 trc; 2 trc in next trc; ch 5, * 2 trc in next trc; trc in next 7 trc, 2 trc in next trc; ch 5; rep from * 6 times more; join in 4th ch of beg ch-4.

Rnd 7: Ch 4, trc in next 3 trc, CL (see Pattern Stitches) over next 2 trc; ch 11, CL over trc just worked and next trc; trc in next 4 trc, ch 3, * trc in next 4 trc, CL over next 2 trc; ch 11, CL over trc just worked and next trc; trc in next 4 trc, ch 3; rep from * 6 times more; join in 4th ch of beg ch-4.

Rnd 8: Ch 4, trc in next 4 trc, ch 5, 5 dc in 6th ch of next ch-11 lp: pineapple base made; ch 5, trc in next 5 trc, ch 1, * trc in next 5 trc, ch 5, 5 dc in 6th ch of next ch-11 lp: pineapple base made; ch 5, trc in next 5 trc, ch 1; rep from * 6 times more; join in 4th ch of beg ch-4: 8 pineapple bases.

Rnd 9: Ch 4, trc in next 4 trc, ch 5, 2 dc in each of next 5 dc; ch 5, * trc in next 10 trc, ch 5, 2 dc in each of next 5 dc; ch 5; rep from * 6 times more; trc in next 5 trc; join in 4th ch of beg ch-4.

Rnd 10: Ch 4, trc in next 4 trc, * ch 4, (dc in next dc, ch 1) 9 times; dc in next dc, ch 4, trc in next 10 trc; rep from * 7 times more, ending last rep without working last 5 trc; join in 4th ch of beg ch-4.

Rnd 11: Ch 4, trc in next 4 trc, * ch 5, sk next ch-4 lp, (sc in next ch-1 sp, ch 5) 9 times; sk next ch-4 lp, trc in next 5 trc, ch 9, trc in next 5 trc; rep from * 7 times more, ending last rep without working last 5 trc; join in 4th ch of beg ch-4.

Rnd 12: Ch 4, trc in next 4 trc, * ch 5, sk next ch-5 lp, (sc in next ch-5 lp, ch 5) 8 times; sk next ch-5 lp, trc in next 5 trc, ch 4, 3 trc in 5th ch of next ch-9 lp; ch 4, trc in next 5 trc; rep from * 7 times more, ending last rep without working last 5 trc; join in 4th ch of beg ch-4.

Rnd 13: Ch 4, trc in next 4 trc, * ch 5, sk next ch-5 lp, (sc in next ch-5 lp, ch 5) 7 times; sk next ch-5 lp, trc in next 5 trc, ch 4, 2 trc in each of next 3 trc; ch 4, trc in next 5 trc; rep from * 7 times more, ending last rep without working last 5 trc; join in 4th ch of beg ch-4.

Rnd 14: Ch 4, trc in next 4 trc, * ch 5, sk next ch-5 lp, (sc in next ch-5 lp, ch 5) 6 times; sk next ch-5 lp, trc in next 5 trc, ch 4, 2 trc in each of next 2 trc; trc in next 2 trc, 2 trc

continued on page 12

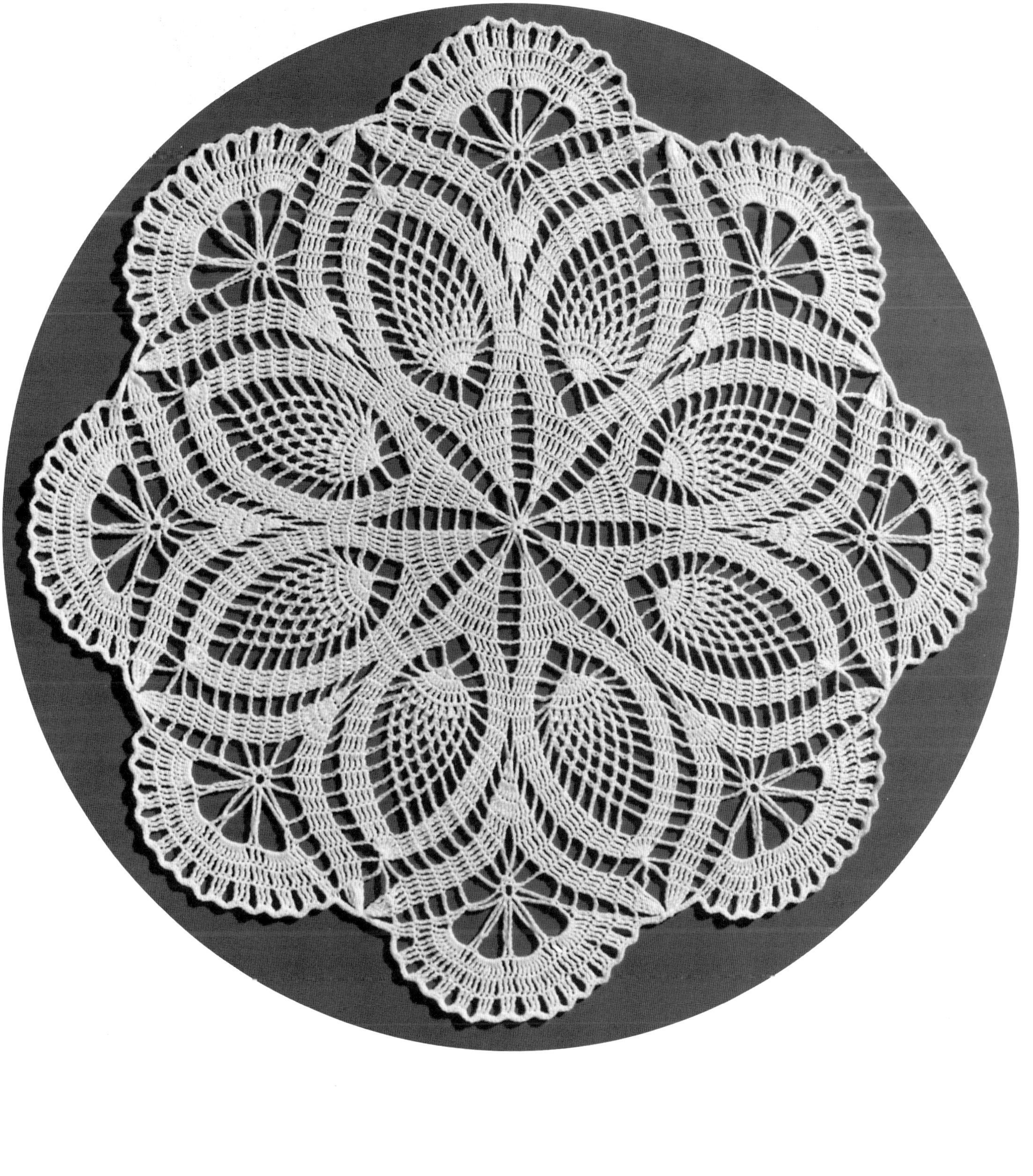

in each of next 2 trc; ch 4, trc in next 5 trc; rep from * 7 times more, ending last rep without working last 5 trc; join in 4th ch of beg ch-4.

Rnd 15: Ch 4, trc in next 4 trc, * ch 5, sk next ch-5 lp, (sc in next ch-5 lp, ch 5) 5 times; sk next ch-5 lp, trc in next 5 trc, ch 4, trc in next 5 trc, ch 9, trc in next 5 trc, ch 4, trc in next 5 trc; rep from * 7 times more, ending last rep without working last 5 trc; join in 4th ch of beg ch-4.

Rnd 16: Ch 4, trc in next 4 trc, * ch 5, sk next ch-5 lp, (sc in next ch-5 lp, ch 5) 4 times; sk next ch-5 lp, trc in next 5 trc, ch 4, trc in next 5 trc, ch 5, sc in next ch-9 lp, ch 5, trc in next 5 trc, ch 4, trc in next 5 trc; rep from * 7 times more, ending last rep without working last 5 trc; join in 4th ch of beg ch-4.

Rnd 17: Ch 4, trc in next 4 trc, * ch 5, sk next ch-5 lp, (sc in next ch-5 lp, ch 5) 3 times; sk next ch-5 lp, trc in next 5 trc, ch 4, trc in next 5 trc, ch 6, sc in next ch-5 lp, in next sc, and in next ch-5 lp close to sc of prev rnd; ch 6, trc in next 5 trc, ch 4, trc in next 5 trc; rep from * 7 times more, ending last rep without working last 5 trc; join in 4th ch of beg ch-4.

Rnd 18: Ch 4, trc in next 4 trc, * ch 5, sk next ch-5 lp, (sc in next ch-5 lp, ch 5) twice; sk next ch-5 lp, trc in next 5 trc, ch 4, trc in next 5 trc, ch 11, sc in next ch-6 lp close to sc of prev rnd and in next sc; ch 3, sk next sc, sc in next sc and in next ch-6 lp; ch 11, trc in next 5 trc, ch 4, trc in next 5 trc; rep from * 7 times more, ending last rep without working last 5 trc; join in 4th ch of beg ch-4.

Rnd 19: Ch 4, trc in next 4 trc, * ch 5, sk next ch-5 lp, sc in next ch-5 lp, ch 5, trc in next 5 trc, ch 4, trc in next 5 trc, ch 7, sc in next ch-11 lp, ch 7, sc in next ch-3 lp, in same lp work (ch 14, sc) 3 times; ch 7, sc in next ch-11 lp, ch 7, trc in next 5 trc, ch 4, trc in next 5 trc; rep from * 7 times more, ending last rep without working last 5 trc; join in 4th ch of beg ch-4.

Rnd 20: Ch 3, trc in next trc: beg CL made; trc in next 3 trc, * sk next 2 ch-5 lps, trc in next 3 trc, CL; ch 4, trc in next 5 trc, ch 7, sc in next ch-7 lp, trc in next sc, (ch 5, 2 trc in next ch-14 lp) 3 times; ch 5, trc in next sc, sc in next ch-7 lp, ch 7, trc in next 5 trc, ch 4, CL; trc in next 3 trc; rep from * 7 times more, ending last rep without working last CL and last 3 trc; join in trc of beg CL.

Rnd 21: Ch 3, keeping last lp of each trc on hook, trc in next 6 trc and in next CL; YO and draw through all 8 lps on hook: beg 8-trc CL made; * ch 5, trc in next 5 trc, ch 6, sc in next ch-7 lp, trc in next trc, (8 trc in next ch-5 lp, trc in next 2 trc) 3 times; 8 trc in next ch-5 lp; trc in next trc, sc in next ch-7 lp, ch 6, trc in next 5 trc, ch 5, 8-trc CL (see Pattern Stitches, page 10) over next CL, next 6 trc, and next CL; rep from * 7 times more, ending last rep without working last 8-trc CL; join in top of beg 8-trc CL.

Rnd 22: Sl st in next 5 chs of next ch-5 lp and in next trc; ch 4, trc in next 4 trc, * ch 5, sc in next ch-6 lp, trc in next 2 trc, ch 1, (trc in next 3 trc, ch 1) 12 times; trc in next 2 trc, sc in next ch-6 lp, ch 5, trc in next 5 trc, ch 6, sc in top of next CL, ch 6, trc in next 5 trc; rep from * 7 times more, ending last rep without working last 5 trc; join in 4th ch of beg ch-4.

Rnd 23: Ch 5 (counts as a tr trc), keeping last lp of each tr trc (see Pattern Stitches, page 10) on hook, tr trc in each of next 4 trc, YO and draw through all 5 lps on hook: beg 5-tr trc CL made; * ch 5, sc in next ch-5 lp, trc in next 2 trc, ch 3, (trc in next 3 trc, ch 3) 12 times; trc in next 2 trc, sc in next ch 5 lp, ch 5, [5-tr trc CL (see Pattern Stitches, page 10) over next 5 trc] twice; rep from * 7 times more, ending last rep without working last 5-tr trc CL; join in top of beg 5-tr trc CL.

Rnd 24: Sl st in next ch-5 lp; ch 1, 6 sc in same lp; * sc in next trc, (5 sc in next ch-3 lp, sk next trc, sc in next trc) 13 times; 6 sc in next ch-5 lp; sc in sp between next 2 CLs, 6 sc in next ch-5 lp; rep from * 7 times more, ending last rep without working last 6 sc; join in first sc.

Finish off and weave in all ends.

Petals

designed by Linda Mershon

Size:

About 18" diameter

Materials:

Bedspread-weight crochet cotton, two 225-yd balls white
Size 6 steel crochet hook, or size required for gauge

Gauge:

8 dc = 1"

Pattern Stitches

Cluster (CL)**:** Keeping last lp of each dc on hook, 2 dc in next st; YO and draw through all 3 lps on hook: CL made.

3-double crochet Cluster (3-dc CL)**:** Keeping last lp of each dc on hook, 3 dc in next st or dc in next 3 sts; YO and draw through all 4 lps on hook: 3-dc CL made.

Joined Cluster (joined CL)**:** keeping last lp of each dc on hook, 2 dc in 4th ch of next ch-5 lp and 2 dc in 2nd ch of next ch-5 lp; YO and draw through all 5 lps on hook: joined CL made.

Instructions

Ch 4, join to form a ring.

Rnd 1: Ch 6 (counts as a dc and a ch-3 lp), * dc in ring, ch 3; rep from * 6 times more; join in 3rd ch of beg ch-6: 8 ch-3 lps.

Rnd 2: Sl st in next ch-3 lp; ch 3 (counts as a dc on this and following rnds), 2 dc in same lp; ch 3, * 3 dc in next ch-3 lp; ch 3; rep from * 6 times more; join in 3rd ch of beg ch-3.

Rnd 3: Ch 3, dc in same ch as joining; ch 1, sk next dc, 2 dc in next dc; ch 3, * 2 dc in next dc; ch 1, sk next dc, 2 dc in next dc; ch 3; rep from * 6 times more; join in 3rd ch of beg ch-3.

Rnd 4: Ch 3, dc in same ch as joining; * dc in next dc, ch 2, dc in next dc, 2 dc in next dc; ch 3, 2 dc in next dc; rep from * 7 times more, ending last rep without working last 2 dc; join in 3rd ch of beg ch-3.

Rnd 5: Ch 3, dc in next 2 dc, ch 3, (dc in next 3 dc, ch 3) 15 times; join in 3rd ch of beg ch-3: sixteen 3-dc groups.

Rnd 6: Ch 3, dc in next 2 dc, * ch 5, dc in next 3 dc, ch 3, dc in next 3 dc; rep from * 7 times more, ending last rep without working last 3 dc; join in 3rd ch of beg ch-3.

Rnd 7: Ch 3, dc in next 2 dc, * ch 2, in 3rd ch of next ch-5 lp work [CL (see Pattern Stitches), ch 2, CL]; ch 2, dc in next 3 dc, ch 3, dc in next 3 dc; rep from * 7 times more, ending last rep without working last 3 dc; join in 3rd ch of beg ch-3.

Rnd 8: Ch 3, dc in next 2 dc, * ch 3, sk next ch-2 sp, CL in next ch-2 sp; ch 3, dc in next 3 dc, ch 3, dc in next 3 dc; rep from * 7 times more, ending last rep without working last 3 dc; join in 3rd ch of beg ch-3.

Rnd 9: Ch 3, dc in next 2 dc, * ch 5, sc in top of next CL, ch 5, dc in next 3 dc, ch 3, dc in next 3 dc; rep from * 7 times more, ending last rep without working last 3 dc; join in 3rd ch of beg ch-3.

Rnd 10: Ch 3, dc in next 2 dc, * ch 4, CL in 3rd ch from hook; 3-dc CL (see Pattern Stitches) in 3rd ch of next ch-5 lp; ch 1, 3-dc CL in 3rd ch of next ch-5 lp; ch 3, CL in 3rd ch from hook; ch 1, dc in next 3 dc, ch 3, dc in next 3 dc; rep from * 7 times more, ending last rep without working last 3 dc; join in 3rd ch of beg ch-3.

Rnd 11: Ch 3, dc in next 2 dc, * ch 4, sk next 2 CLs, in next ch work (3-dc CL, ch 4, 3-dc CL); ch 4, dc in next 3 dc, ch 5, dc in next 3 dc; rep from * 7 times more, ending last rep without working last 3 dc; join in 3rd ch of beg ch-3.

Rnd 12: Ch 3, dc in next 2 dc, * ch 5, sk next ch-4 lp, sc in next ch-4 lp, ch 5, dc in next 3 dc, ch 3, dc in 3rd ch of next ch-5 lp, ch 3, dc in next 3 dc; rep from * 7 times more, ending last rep without working last 3 dc; join in 3rd ch of beg ch-3.

Rnd 13: Ch 3, dc in next 2 dc, * ch 5, CL in next sc; ch 5, dc in next 3 dc, ch 3, 3 dc in next dc; ch 3, dc in next 3 dc; rep from * 7 times more, ending last rep without working last 3 dc; join in 3rd ch of beg ch-3.

Rnd 14: Ch 3, dc in next 2 dc, * ch 4, joined CL (see Pattern Stitches); ch 4, dc in next 3 dc, ch 3, 2 dc in next dc; ch 1, sk next dc, 2 dc in next dc; ch 3, dc in next 3 dc; rep from * 7 times more, ending last rep without working last 3 dc; join in 3rd ch of beg ch-3.

Rnd 15: Ch 3, dc in next 2 dc, * ch 5, dc in next 3 dc, ch 5, 2 dc in next dc; dc in next dc, ch 2, dc in next dc, 2 dc in next dc; ch 5, dc in next 3 dc; rep from * 7 times more, ending last rep without working last 3 dc; join in 3rd ch of beg ch-3.

continued

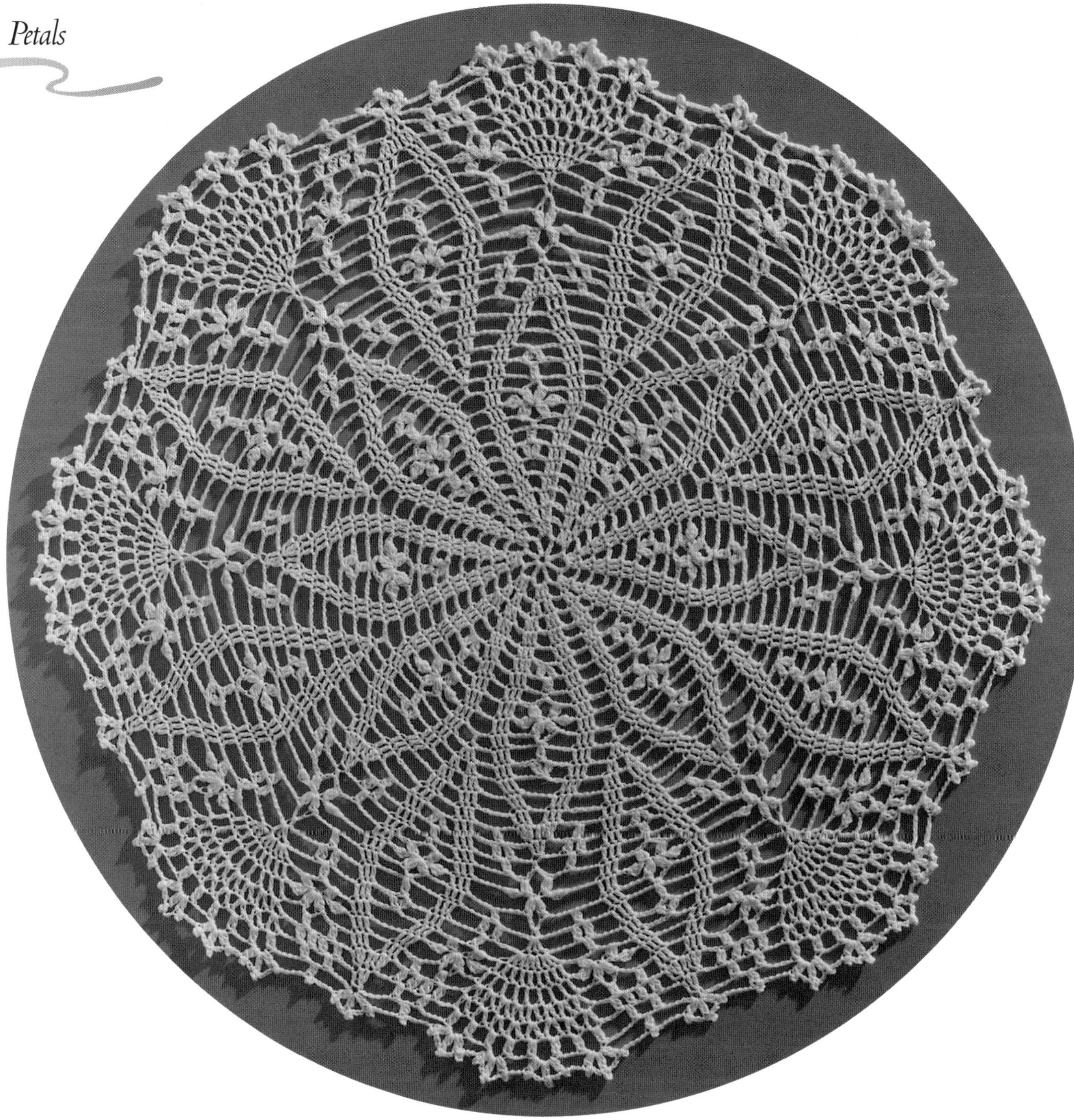

Rnd 16: Ch 3, dc in next 2 dc, ch 3, * dc in next 3 dc, ch 6, dc in next 3 dc, ch 3; rep from * around, ending last rep without working last 3 dc and last ch 3; join in 3rd ch of beg ch-3.

Rnd 17: Ch 3, dc in next 2 dc, * ch 1, dc in next 3 dc, ch 7, dc in next 3 dc, ch 5, dc in next 3 dc, ch 7, dc in next 3 dc; rep from * 7 times more, ending last rep without working last 3 dc; join in 3rd ch of beg ch-3.

Rnd 18: Ch 2, CL over next 2 dc; * dc in next ch-l sp, 3-dc CL over next 3 dc; ch 5, CL in 4th ch of next ch-7 lp; ch 5, dc in next 3 dc, ch 2, in 3rd ch of next ch-5 lp work (CL, ch 2, CL); ch 2, dc in next 3 dc, ch 5, CL in 4th ch of next ch-7 lp; ch 5, 3-dc CL over next 3 dc; rep from * 7 times more, ending last rep without working last 3-dc CL; join in top of first CL.

Rnd 19: Ch 2, CL over next dc and next CL; * † ch 3, CL in 5th ch of next ch-5 lp; ch 1, CL in first ch of next ch-5 lp; ch 3 †, dc in next 3 dc, ch 3, sk next ch-2 sp, CL in next ch-2 sp; ch 3, dc in next 3 dc; rep from † to † once; 3-dc CL over next CL, next dc, and next CL; rep from * 7 times more, ending last rep without working last 3-dc CL; join in top of first CL.

Rnd 20: Ch 1, sc in same CL as joining; * † ch 5, CL in next ch-l sp; ch 5 †, dc in next 3 dc, ch 5, sc in top of next CL, ch 5, dc in next 3 dc; rep from † to † once; sc in top of next CL; rep from * 7 times more, ending last rep without working last sc; join in first sc.

Rnd 21: Sl st in next 3 chs of next ch-5 lp; ch 3, CL in same ch as last sl st made; ch 3, CL in 3rd ch from hook; ch 5, dc in next 3 dc, * † ch 4, CL in 3rd ch from hook; 3-dc CL in 3rd ch of next ch-5 lp; ch 1, 3-dc CL in 3rd ch of next ch-5 lp; ch 3, CL in 3rd ch from hook; ch 1, dc in next 3 dc, ch 8, CL in 3rd ch from hook; sk next ch-5 lp, 3-dc CL in 3rd ch of next ch-5 lp; ch 1 †, 3-dc CL in 3rd ch of next ch-5 lp; ch 3, CL in 3rd ch from hook; ch 5, dc in next 3 dc; rep from * 6 times more, then rep from † to † once; join in 3rd ch of beg ch-3.

Rnd 22: 2 sl sts in side of next CL; sl st in next 5 chs and in next dc; ch 3, dc in next 2 dc, * ch 4, in next ch work (3-dc CL, ch 4, 3-dc CL); ch 4, dc in next 3 dc, ch 9, in next ch work (3-dc CL, ch 4, 3-dc CL); ch 9, dc in next 3 dc; rep from * 7 times more, ending last rep without working last 3 dc; join in 3rd ch of beg ch-3.

Rnd 23: Ch 3, dc in next 2 dc, * ch 5, sk next ch-4 lp, sc in next ch-4 lp, ch 5, dc in next 3 dc, ch 5, CL in 5th ch of next ch-9 lp; ch 7, sc in next ch-4 lp, ch 7, CL in 5th ch of next ch-9 lp; ch 5, dc in next 3 dc; rep from * 7 times more, ending last rep without working last 3 dc; join in 3rd ch of beg ch-3.

Rnd 24: Ch 3, dc in next 2 dc, * ch 5, CL in next sc; ch 5, dc in next 3 dc, ch 5, CL in 5th ch of next ch-5 lp; ch 1, CL in first ch of next ch-7 lp; ch 5, 3 dc in next sc; ch 5, CL in 7th ch of next ch-7 lp; ch 1, CL in first ch of next ch-5 lp; ch 5, dc in next 3 dc; rep from * 7 times more, ending last rep without working last 3 dc; join in 3rd ch of beg ch-3.

Rnd 25: Ch 3, dc in next 2 dc, * ch 4, joined CL; ch 4, dc in next 3 dc, ch 5, CL in next ch-l sp; ch 5, 3 dc in each of next 3 dc; ch 5, CL in next ch-l sp; ch 5, dc in next 3 dc; rep from * 7 times more, ending last rep without working last 3 dc; join in 3rd ch of beg ch-3.

Rnd 26: Ch 3, dc in next 2 dc, * ch 5, dc in next 3 dc, † ch 4, CL in 3rd ch from hook; 3-dc CL in 3rd ch of next ch-5 lp; ch 1, 3-dc CL in 3rd ch of next ch-5 lp; ch 3, CL in 3rd ch from hook; ch 1 †, dc in next dc, (ch 1, dc in next dc) 8 times; rep from † to † once; dc in next 3 dc; rep from * 7 times more, ending last rep without working last 3 dc; join in 3rd ch of beg ch-3.

Rnd 27: Ch 3, dc in next 2 dc, * ch 3, dc in next 3 dc, ch 7, in next ch work (3-dc CL, ch 4, 3-dc CL); ch 3, dc in next dc, (ch 2, dc in next dc) 8 times; ch 3, in next ch work (3-dc CL, ch 4, 3-dc CL); ch 7, dc in next 3 dc; rep from * 7 times more, ending last rep without working last 3 dc; join in 3rd ch of beg ch-3.

Rnd 28: Ch 3, dc in next 2 dc, * ch 1, dc in next 3 dc, ch 3, CL in 4th ch of next ch-7 lp; ch 3, CL in next ch-4 lp; ch 3, in each of next 8 ch-2 sps work (dc, ch 2, dc); ch 3, CL in next ch-4 lp; ch 3, CL in 4th ch of next ch-7 lp; ch 3, dc in next 3 dc; rep from * 7 times more, ending last rep without working last 3 dc; join in 3rd ch of beg ch-3.

Rnd 29: Ch 2, CL over next 2 dc; * dc in next ch-l sp, 3-dc CL over next 3 dc; † ch 3, CL in 3rd ch of next ch-3 lp; ch 1, sk next CL, CL in next ch; ch 1, sk next ch, CL in next ch; ch 1, sk next CL, CL in next ch; ch 3 †, in each of next 8 ch-2 sps work (dc, ch 2, dc); rep from † to † once; 3-dc CL over next 3 dc; rep from * 7 times more, ending last rep without working last 3-dc CL; join in top of first CL.

Rnd 30: Ch 2, CL over next dc and next CL; * † ch 5, CL in next ch-l sp; ch 3, sk next ch-l sp, CL in next ch-l sp; ch 5 †, CL in next ch-2 sp; (ch 3, CL in next ch-2 sp) 7 times; rep from † to † once; 3-dc CL over next CL, next dc, and next CL; rep from * 7 times more, ending last rep without working last 3-dc CL; join in top of first CL.

Rnd 31: Ch 2, dc in same CL as joining; ch 3, sl st in dc just made; in same CL work (ch 3, CL, ch 3, sl st in CL just made) twice; * † ch 5, sk next ch-5 lp, CL in 2nd ch of next ch-3 lp; ch 3, sl st in CL just made; ch 5, sk next ch-5 lp, †† in next ch-3 lp work (CL, ch 3, sl st in CL just made; ch 3) 3 times; sk next ch-3 lp ††; rep from †† to †† twice more; in next ch-3 lp work (CL, ch 3, sl st in CL just made; ch 3) twice; in same lp work (CL, ch 3, sl st in CL just made); ch 5, sk next ch-5 lp, CL in 2nd ch of next ch-3 lp; ch 3, sl st in CL just made; ch 5 †, sk next CL, in next CL work (CL, ch 3, sl st in CL just made); in same CL work (ch 3, CL, ch 3, sl st in CL just made) twice; rep from * 6 times more, then rep from † to † once; join in first dc.

Finish off and weave in all ends.

Summer Nights

designed by Jo Ann Maxwell

Size:

About 17" diameter

Materials:

Bedspread-weight crochet cotton, one 255-yd ball white
Size 6 steel crochet hook, or size required for gauge

Gauge:

8 dc = 1"

Instructions

Ch 5, join to form a ring.

Rnd 1: Ch 3 (counts as a dc on this and following rnds), 19 dc in ring; join in 3rd ch of beg ch-3; 20 dc.

Rnd 2: Ch 6 (counts as a trc and a ch-2 sp), * trc in next dc, ch 2; rep from * 18 times more; join in 4th ch of beg ch-6.

Rnd 3: Ch 3, dc in each ch and in each trc; join in 3rd ch of beg ch-3: 60 dc.

Rnd 4: Ch 3, dc in same ch as joining; ch 3, sk next 2 dc, * 2 dc in next dc; ch 3, sk next 2 dc; rep from * around; join in 3rd ch of beg ch-3: 20 ch-3 lps.

Rnd 5: Sl st in next dc and in next 2 chs and in same lp; ch 7, sl st in 4th ch from hook: beg picot made; ch 5, * trc in next ch-3 lp; ch 4, sl st in trc just made: picot made; ch 5; rep from * around; join in base of beg picot.

Rnd 6: Sl st in next 3 chs of next ch-5 lp; ch 3, in same ch as last sl st made work (dc, ch 2, 2 dc): beg shell made; ch 3, * in 3rd ch of next ch-5 lp work (2 dc, ch 2, 2 dc): shell made; ch 3; rep from * around; join in 3rd ch of beg ch-3: 20 shells.

Rnd 7: Sl st in next dc and in next ch-2 sp; ch 1, sc in same sp; ch 9, * sc in ch-2 sp of next shell, ch 9; rep from * around; join in first sc.

Rnd 8: Sl st in next 2 chs of next ch-9 lp; ch 3, dc in same ch as last sl st made; * (dc in next ch, 2 dc in next ch) 3 times; 2 dc in 2nd ch of next ch-9 lp; 2 dc in next ch; rep from * around, ending last rep without working last 2 dc; join in 3rd ch of beg ch-3.

Rnd 9: Ch 1, sc in same ch as joining; picot; * ch 4, sk next 4 dc, in next dc work (2 dc, ch 2, 2 dc): shell made; ch 4, sk next 5 dc, sc in next dc, picot; rep from * around, ending last rep without working last sc and picot; join in first sc.

Rnd 10: Sl st in next 4 chs, in next 2 dc, and in next ch-2 sp; ch 1, sc in same sp; picot; ch 13, * sc in ch-2 sp of next shell, picot; ch 13; rep from * around; join in first sc.

Rnd 11: Sl st in next 7 chs of next ch-13 lp; ch 1, in same ch as last sl st made work (sc, hdc, dc, 3 trc, dc, hdc, sc); ch 11, * in 7th ch of next ch-13 lp work (sc, hdc, dc, 3 trc, dc, hdc, sc); ch 11; rep from * around; join in first sc.

Rnd 12: Sl st in next hdc, in next dc, and in next 2 trc; ch 1, sc in same trc as last sl st made; picot; * ch 7, sc in 6th ch of next ch-11 lp, picot; ch 7, sk next 4 sts, sc in next trc, picot; rep from * around, ending last rep without working last sc and picot; join in first sc.

Rnd 13: Sl st in next 4 chs of next ch-7 lp and in same lp; ch 1, sc in same lp; ch 8, * sc in next ch-7 lp, ch 8; rep from * around; join in first sc.

Rnd 14: Sl st in next 4 chs of next ch-8 lp and in same lp; ch 1, sc in same lp; ch 9, * sc in next ch-8 lp, ch 9; rep from * around; join in first sc.

Rnd 15: Sl st in next 5 chs of next ch-9 lp and in same lp; ch 1, sc in same lp; ch 9, * sc in next ch-9 lp, ch 9; rep from * around; join in first sc.

Rnds 16 through 20: Rep Rnds 8 through 12. At end of Rnd 20, finish off and weave in ends.

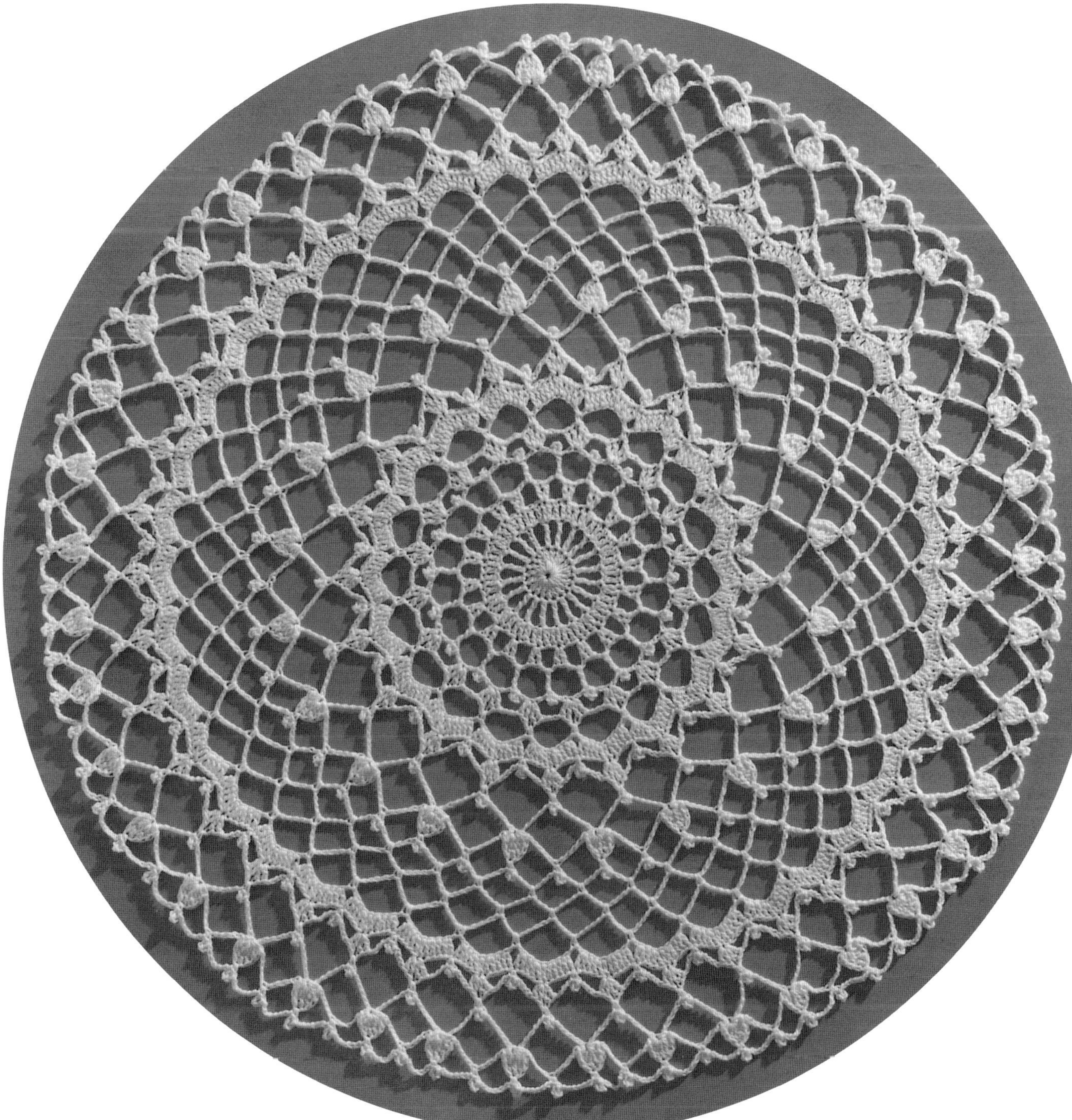

Swinging Bells

designed by Jo Ann Maxwell

Size:

About 14" diameter

Materials:

Bedspread-weight crochet cotton, one 225-yd ball white
Size 6 steel crochet hook, or size required for gauge

Gauge:

8 dc = 1"

Instructions

Ch 5; join to form a ring.

Rnd 1: Ch 3 (counts as a dc on this and following rnds), 19 dc in ring; join in 3rd ch of beg ch-3: 20 dc.

Rnd 2: Ch 4 (counts as a dc and a ch-1 sp), * dc in next dc, ch 1; rep from * 18 times more; join in 3rd ch of beg ch-4.

Rnd 3: Ch 1, sc in same ch as joining; ch 2, * sc in next dc, ch 2; rep from * around; join in first sc.

Rnd 4: Sl st in next ch-2 sp; ch 3, in same sp work (dc, ch 3, sl st in st just made: picot made; dc); ch 1, * in next ch-2 sp work (2 dc, ch 3, sl st in st just made: picot made; dc); ch 1; rep from * around; join in 3rd ch of beg ch-3.

Rnd 5: Sl st in next 2 dc and in next ch-1 sp; ch 1, sc in same sp; ch 6, * sc in next ch-1 sp, ch 6; rep from * around; join in first sc.

Rnd 6: Sl st in next 3 chs of next ch-6 lp and in same lp; ch 1, sc in same lp; ch 5, * sc in next ch-6 lp, ch 5; rep from * around; join in first sc.

Rnd 7: Ch 5 (counts as a dc and a ch-2 sp), dc in same sc as joining: beg V-st made; ch 2, sc in next ch-5 lp, ch 2, * in next sc work (dc, ch 2, dc): V-st made; ch 2, sc in next ch-5 lp, ch 2; rep from * around; join in 3rd ch of beg ch-5: 20 V-sts.

Rnd 8: Sl st in next ch-2 sp; ch 3, 4 dc in same sp; * ch 3 (counts as a dc) turn; dc in first dc and in next 3 dc; 2 dc in next dc: 7-dc group made; ch 1, turn; sc in next 4 dc, picot; sc in next 3 dc: bell made; ch 2, sc in ch-2 sp of next V-st, ch 5, 5 dc in ch-2 sp of next V-st; rep from * around, ending last rep without working last 5 dc; join in 3rd ch of beg ch-3 (side of first bell): 10 bells. Finish off.

Hold doily with right side facing you; join thread in right-hand (unjoined) corner of any bell.

Rnd 9: Ch 15, * sl st in right-hand corner of next bell; ch 15; rep from * around; join in joining sl st of prev rnd.

Rnd 10: Sl st in next 2 chs of next ch-15 lp; ch 3, dc in same ch as last sl st made; (dc in next ch, 2 dc in next ch) 6 times; * 2 dc in 2nd ch of next ch-15 lp; (dc in next ch, 2 dc in next ch) 6 times; rep from * around; join in 3rd ch of beg ch-3.

Rnd 11: Ch 1, sc in same ch as joining; ch 4, sk next 3 dc, * sc in next dc, ch 4, sk next 3 dc; rep from * around; join in first sc.

Rnd 12: Sl st in next 2 chs of next ch-4 lp and in same lp; ch 1, sc in same lp; ch 4, * sc in next ch-4 lp, ch 4; rep from * around; join in first sc.

Rnd 13: Rep Rnd 12.

Rnd 14: Sl st in next 2 chs of next ch-4 lp and in same lp; ch 3, in same lp work (dc, picot; dc); ch 2, * in next ch-4 lp work (2 dc, picot; dc); ch 2; rep from * around; join in 3rd ch of beg ch-3.

Rnd 15: Sl st in next 2 dc and in next ch-2 sp; ch 1, sc in same sp; ch 6, * sc in next ch-2 sp, ch 6; rep from * around; join in first sc.

Rnd 16: Rep Rnd 6.

Rnd 17: Rep Rnd 7. At end of rnd: 50 V-sts.

Note: Bells on following rnd are worked in opposite direction of bells on Rnd 8.

Rnd 18: Sl st in next 2 chs of next ch-2 sp and in same sp; ch 3, turn; 4 dc in same sp; * ch 3 (counts as a dc), turn; dc in first dc and in next 3 dc; 2 dc in next dc: 7-dc group made; ch 1, turn; sc in next 4 dc, picot; sc in next 3 dc: bell made; ch 2, sc in ch-2 sp of next V-st, ch 5, 5 dc in ch-2 sp of next V-st; rep from * around, ending last rep without working last 5 dc; join in 3rd ch of beg ch-3 (side of first bell): 25 bells. Finish off.

Hold doily with right side facing you; join thread in left-hand (unjoined) corner of any bell.

Rnd 19: Ch 15, * sl st in left-hand corner of next bell; ch 15; rep from * around; join in joining sl st of prev rnd.

Rnds 20 and 21: Rep Rnds 10 and 11. At end of Rnd 21, finish off and weave in all ends.

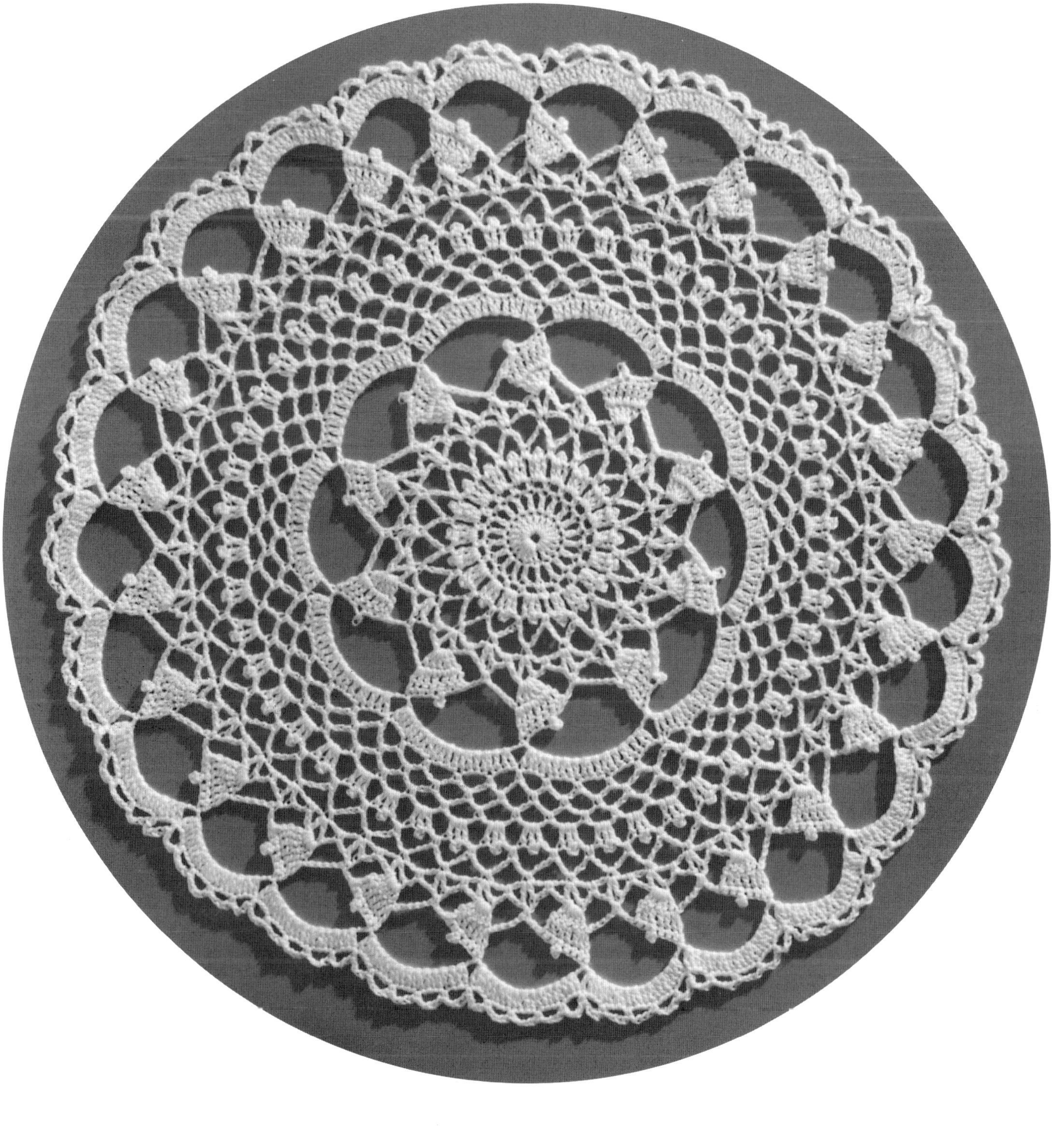

Candlelight

designed by Jo Ann Maxwell

Size:

About 27" diameter

Materials:

Bedspread-weight crochet cotton, three 225-yd balls cream

Size 7 steel crochet hook, or size required for gauge

Gauge:

9 dc = 1"

Pattern Stitches

Cluster (CL)**:** Keeping last lp of each dc on hook, dc in next 2 dc, YO and draw through 3 lps on hook: CL made.

3-dc Cluster (3-dc CL)**:** Keeping last lp of each dc on hook, dc in next 3 sts, YO and draw through all 4 lps on hook: 3-dc CL made.

4-dc Cluster (4-dc CL)**:** Keeping last lp of each dc on hook, dc in next 4 sts, YO and draw through all 5 lps on hook: 4-dc CL made.

Instructions

Ch 4, join to form a ring.

Rnd 1: Ch 3 (counts as a dc on this and following rnds), 19 dc in ring; join in 3rd ch of beg ch-3: 20 dc.

Rnd 2: Ch 1, sc in same ch as joining; ch 11, sk next dc, * sc in next dc, ch 11, sk next dc; rep from * 8 times more; join in first sc: 10 ch-11 lps.

Rnd 3: Sl st in next 5 chs of next ch-11 lp; ch 3, dc in same ch as last sl st made; ch 3, sk next ch, 2 dc in next ch; ch 3, * 2 dc in 5th ch of next ch-11 lp; ch 3, sk next ch, 2 dc in next ch; ch 3; rep from * 8 times more; join in 3rd ch of beg ch-3.

Rnd 4: Sl st in next dc and in next ch-3 lp; ch 1, sc in same lp; ch 3, sl st in st just made: picot made; ch 9, * sk next ch-3 lp, sc in next ch-3 lp; ch 3, sl st in sc just made: picot made; ch 9; rep from * 8 times more; join in first sc.

Rnd 5: Sl st in next 2 chs of next ch-9 lp; ch 3, dc in same ch as last sl st made; (ch 2, sk next 2 chs, 2 dc in next ch) twice; ch 3, sk next picot, * 2 dc in 2nd ch of next ch-9 lp; (ch 2, sk next 2 chs, 2 dc in next ch) twice; ch 3, sk next picot; rep from * 8 times more; join in 3rd ch of beg ch-3.

Rnd 6: Sl st in next dc, in next 2 chs, and in next dc; ch 3, dc in same dc as last sl st made; ch 3, 2 dc in next dc; ch 4, sc in next ch-3 lp, picot; ch 4, sk next 2 dc, * 2 dc in next dc; ch 3, 2 dc in next dc; ch 4, sc in next ch-3 lp, picot; ch 4, sk next 2 dc; rep from * 8 times more; join in 3rd ch of beg ch-3.

Rnd 7: Sl st in next dc and in next ch-3 lp; ch 3, 8 dc in same lp: beg candle base made; ch 7, * sk next 2 ch-4 lps, 9 dc in next ch-3 lp: candle base made; ch 7; rep from * 8 times more; join in 3rd ch of beg ch-3: 10 candle bases.

Rnd 8: Ch 3, dc in same ch as joining; * dc in next dc, ch 3, sk next 2 dc, sc in next dc, picot; ch 3, sk next 2 dc, dc in next dc, 2 dc in next dc; ch 2, sc in 4th ch of next ch-7 lp, picot; ch 2, 2 dc in next dc; rep from * 9 times more, ending last rep without working last 2 dc; join in 3rd ch of beg ch-3.

Rnd 9: Ch 3, dc in next 2 dc, * ch 2, sc in next ch-3 lp, ch 7, sc in next ch-3 lp, ch 2, dc in next 3 dc, ch 5, dc in next 3 dc; rep from * 9 times more, ending last rep without working last 3 dc; join in 3rd ch of beg ch-3.

Rnd 10: Ch 3, dc in next 2 dc, * ch 2, sk next ch of next ch-7 lp, sc in next 2 chs, 3 hdc in next ch; sc in next 2 chs, ch 2, sk next ch, dc in next 3 dc, ch 3, sc in 3rd ch of next ch-5 lp, picot; ch 3, dc in next 3 dc; rep from * 9 times more, ending last rep without working last 3 dc; join in 3rd ch of beg ch-3.

Rnd 11: Ch 3, dc in next 2 dc, * ch 2, sk next 2 sc and next hdc; sc in next hdc, picot; ch 2, dc in next 3 dc, ch 9, dc in next 3 dc; rep from * 9 times more, ending last rep without working last 3 dc; join in 3rd ch of beg ch-3.

Rnd 12: Ch 3, dc in next 2 dc, * ch 3, dc in next 3 dc, ch 5, 2 dc in 4th ch of next ch-9 lp; ch 3, sk next ch, 2 dc in next ch; ch 5, dc in next 3 dc; rep from * 9 times more, ending last rep without working last 3 dc; join in 3rd ch of beg ch-3.

Rnd 13: Ch 3, dc in next 2 dc, * ch 1, dc in next 3 dc, † ch 3, sc in 3rd ch of next ch-5 lp, picot; ch 3 †, 2 dc in first ch of next ch-3 lp; ch 3, sk next ch, 2 dc in next ch; rep from † to † once; dc in next 3 dc; rep from * 9 times more, ending last rep without working last 3 dc; join in 3rd ch of beg ch-3.

Rnd 14: Ch 3, * CL (see Pattern Stitches) over next 2 dc; sk next ch-1 sp, CL over next 2 dc; dc in next dc, ch 9, sk next 2 ch-3 lps, 2 dc in first ch of next ch-3 lp; ch 3, sk next ch, 2 dc in next ch; ch 9, sk next 2 ch-3 lps, dc in next dc; rep from * 9 times more, ending last rep without working last dc; join in 3rd ch of beg ch-3.

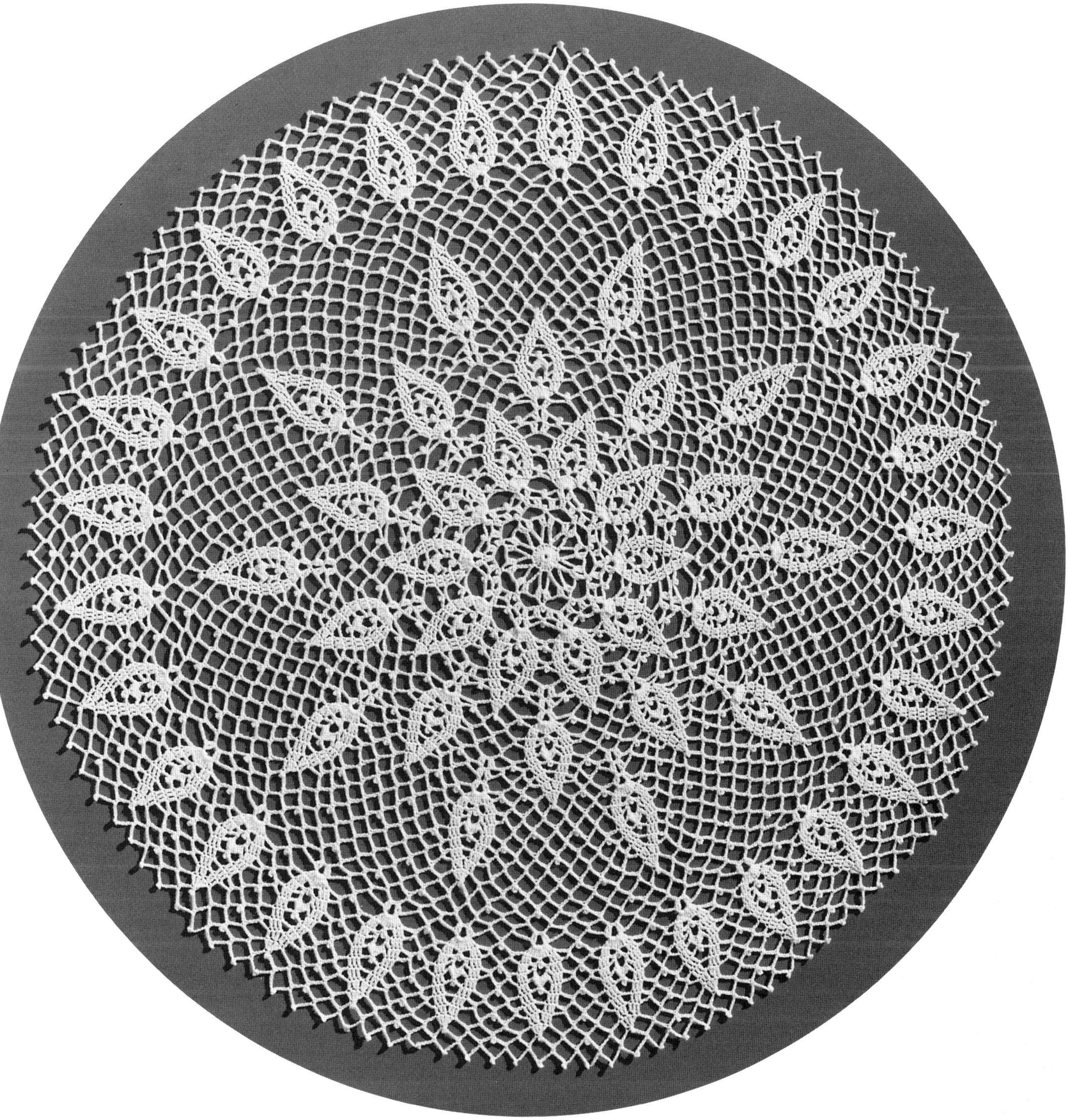

Rnd 15: Ch 2, 3-dc CL (see Pattern Stitches) over next 3 sts; picot; * † ch 7, sc in 5th ch of next ch-9 lp, picot; ch 3, 9 dc in next ch-3 lp: candle base made; ch 3, sc in 5th ch of next ch-9 lp, picot †; ch 7, 4-dc CL (see Pattern Stitches) over next 3 sts; picot; rep from * 8 times more, then rep from † to † once; ch 2, trc in top of first 3-dc CL.

Rnd 16: Ch 1, sc in last trc made; ch 7, sc in 4th ch of next ch-7 lp, ch 7, 2 dc in next dc; * † dc in next dc, ch 3, sk next 2 dc, sc in next dc, picot; ch 3, sk next 2 dc, dc in next dc, 2 dc in next dc; ch 7 †, (sc in 4th ch of next ch-7 lp, ch 7) twice; 2 dc in next dc; rep from * 8 times more, then rep from † to † once; join in first sc.

Rnd 17: Sl st in next 4 chs of next ch-7 lp; ch 1, sc in same ch as last sl st made; picot; * † ch 7, sc in 4th ch of next ch-7 lp, ch 3, dc in next 3 dc, ch 2, sc in 2nd ch of next ch-3 lp, ch 7, sc in 2nd ch of next ch-3 lp, ch 2, dc in next 3 dc, ch 3, sc in 4th ch of next ch-7 lp, ch 7 †, sc in 4th ch of next ch-7 lp, picot; rep from * 8 times more, then rep from † to † once; join in first sc.

continued

Rnd 18: Sl st in next 4 chs of next ch-7 lp; ch 1, sc in same ch as last sl st made; ch 7, dc in next 3 dc, * † ch 2, sk next ch of next ch-7 lp, sc in next 2 chs, 3 hdc in next ch; sc in next 2 chs, ch 2, dc in next 3 dc, ch 7 †, (sc in 4th ch of next ch-7 lp, ch 7) twice; dc in next 3 dc; rep from * 8 times more, then rep from † to † once; sc in 4th ch of next ch-7 lp, ch 7; join in first sc.

Rnd 19: Sl st in next 4 chs of next ch-7 lp; ch 1, sc in same ch as last sl st made; picot; ch 5, dc in next 3 dc, * † ch 2, sk next 2 sc and next hdc; sc in next hdc, picot; ch 2, dc in next 3 dc, ch 5, sc in 4th ch of next ch-7 lp, picot; ch 5, 2 dc in 3rd ch of next ch-7 lp; ch 3, sk next ch, 2 dc in next ch; ch 5 †, sc in 4th ch of next ch-7 lp, picot; ch 5, dc in next 3 dc; rep from * 8 times more, then rep from † to † once; join in first sc.

Rnd 20: Sl st in next 3 chs of next ch-5 lp; ch 1, sc in same ch as last sl st made; ch 5, dc in next 3 dc, * † ch 3, dc in next 3 dc, ch 5, sc in 3rd ch of next ch-5 lp, ch 7, sc in 3rd ch of next ch-5 lp, ch 3, 2 dc in first ch of next ch-3 lp; ch 3, sk next ch, 2 dc in next ch; ch 3, sc in 3rd ch of next ch-5 lp, ch 7 †, sc in 3rd ch of next ch-5 lp, ch 5, dc in next 3 dc; rep from * 8 times more, then rep from † to † once; join in first sc.

Rnd 21: Sl st in next 3 chs of next ch-5 lp; ch 1, sc in same ch as last sl st made; ch 3, dc in next 3 dc, * † ch 1, dc in next 3 dc, ch 3, sc in 3rd ch of next ch-5 lp, ch 5, sc in 4th ch of next ch-7 lp, ch 7, sk next ch-3 lp, 2 dc in first ch of next ch-3 lp; ch 3, sk next ch, 2 dc in next ch; ch 7, sc in 4th ch of next ch-7 lp, ch 5 †, sc in 3rd ch of next ch-5 lp, ch 3, dc in next 3 dc; rep from * 8 times more, then rep from † to † once; join in first sc.

Rnd 22: Sl st in next 3 chs of next ch-3 lp and in next dc; ch 3, * CL; sk next ch-1 sp, CL; dc in next dc, ch 7, sc in 3rd ch of next ch-5 lp, picot; ch 7, sc in 4th ch of next ch-7 lp, picot; ch 3, 9 dc in next ch-3 lp: candle base made; ch 3, sc in 4th ch of next ch-7 lp, picot; ch 7, sc in 3rd ch of next ch-5 lp, picot; ch 7, dc in next dc; rep from * 9 times more, ending last rep without working last dc; join in 3rd ch of beg ch-3.

Rnd 23: Ch 2, 3-dc CL; picot; * † (ch 7, sc in 4th ch of next ch-7 lp) twice; ch 5, 2 dc in next dc; dc in next dc, ch 3, sk next 2 dc, sc in next dc, picot; ch 3, sk next 2 dc, dc in next dc, 2 dc in next dc; ch 5, (sc in 4th ch of next ch-7 lp, ch 7) twice †; 4-dc CL; picot; rep from * 8 times more, then rep from † to † once; join in top of first 3-dc CL.

Rnd 24: Sl st in next 4 chs of next ch-7 lp; ch 1, sc in same ch as last sl st made; ch 7, sc in 4th ch of next ch-7 lp, ch 7, sc in 3rd ch of next ch-5 lp, ch 3, dc in next 3 dc, * † ch 2, sc in 2nd ch of next ch-3 lp, ch 7, sc in 2nd ch of next ch-3 lp, ch 2, dc in next 3 dc, ch 3, sc in 3rd ch of next ch-5 lp †, (ch 7, sc in 4th ch of next ch-7 lp) 4 times; ch 7, sc in 3rd ch of next ch-5 lp, ch 3, dc in next 3 dc; rep from * 8 times more, then rep from † to † once; (ch 7, sc in 4th ch of next ch-7 lp) twice; ch 7; join in first sc.

Rnd 25: Sl st in next 4 chs of next ch-7 lp; ch 1, sc in same ch as last sl st made; picot; ch 7, sc in 4th ch of next ch-7 lp, ch 7, dc in next 3 dc, * † ch 2, sk next ch of next ch-7 lp, sc in next 2 chs, 3 hdc in next ch; sc in next 2 chs, ch 2, dc in next 3 dc, (ch 7, sc in 4th ch of next ch-7 lp) twice; ch 7, sc in 4th ch of next ch-7 lp, picot; ch 7, sc in 4th ch of next ch-7 lp, ch 7 †, sc in 4th ch of next ch-7 lp, picot; ch 7, sc in 4th ch of next ch-7 lp, ch 7, dc in next 3 dc; rep from * 8 times more, then rep from † to † once; join in first sc.

Rnd 26: Sl st in next 4 chs of next ch-7 lp; ch 1, sc in same ch as last sl st made; ch 7, sc in 4th ch of next ch-7 lp, picot; ch 5, dc in next 3 dc, * † ch 2, sk next 2 sc and next hdc; sc in next hdc, picot; ch 2, dc in next 3 dc, ch 5, sc in 4th ch of next ch-7 lp, picot †; (ch 7, sc in 4th ch of next ch-7 lp) 5 times; picot in last sc made; ch 5, dc in next 3 dc; rep from * 8 times more, then rep from † to † once; (ch 7, sc in 4th ch of next ch-7 lp) 3 times; ch 7; join in first sc.

Rnd 27: Sl st in next 4 chs of next ch-7 lp; ch 1, sc in same ch as last sl st made; ch 7, sc in 3rd ch of next ch-5 lp, ch 5, dc in next 3 dc, * † ch 3, dc in next 3 dc, ch 5, sc in 3rd ch of next ch-5 lp, (ch 7, sc in 4th ch of next ch-7 lp) 3 times; picot in last sc made †; (ch 7, sc in 4th ch of next ch-7 lp) twice; ch 7, sc in 3rd ch of next ch-5 lp, ch 5, dc in next 3 dc; rep from * 8 times more, then rep from † to † once; ch 7, sc in 4th ch of next ch-7 lp, ch 7; join in first sc.

Rnd 28: Sl st in next 4 chs of next ch-7 lp; ch 1, sc in same ch as last sl st made; ch 7, sc in 3rd ch of next ch-5 lp, ch 3, dc in next 3 dc, * † ch 1, dc in next 3 dc, ch 3, sc in 3rd ch of next ch-5 lp †, (ch 7, sc in 4th ch of next ch-7 lp) 6 times; ch 7, sc in 3rd ch of next ch-5 lp; ch 3, dc in next 3 dc; rep from * 8 times more, then rep from † to † once; (ch 7, sc in 4th ch of next ch-7 lp) 5 times; ch 7; join in first sc.

Rnd 29: Sl st in next 4 chs of next ch-7 lp; ch 1, sc in same ch as last sl st made; ch 7, * † dc in next dc, CL; sk next ch-1 sp, CL; dc in next dc, (ch 7, sc in 4th ch of next ch-7 lp) twice; picot in last sc made; (ch 7, sc in 4th ch of next ch-7 lp) 4 times; picot in last sc made; ch 7 †, sc in 4th ch of next ch-7 lp, ch 7; rep from * 8 times more, then rep from † to † once; join in first sc.

Rnd 30: Sl st in next 4 chs of next ch-7 lp; ch 1, sc in same ch as last sl st made; * ch 7, 4-dc CL; picot; (ch 7, sc in 4th ch of next ch-7 lp) 8 times; rep from * around, ending last rep without working last sc; join in first sc.

Rnd 31: Sl st in next 4 chs of next ch-7 lp; ch 1, sc in same lp as last sl st made; * † (ch 7, sc in 4th ch of next ch-7 lp) twice; picot in last sc made †; rep from † to † 3 times more; ch 7, sc in 4th ch of next ch-7 lp; rep from * around, ending last rep without working last sc; join in first sc: 90 ch-7 lps.

Rnd 32: Sl st in next 4 chs of next ch-7 lp; ch 1, sc in same ch as last sl st made; ch 7, (sc in 4th ch of next ch-7 lp, ch 7) 89 times; join in first sc.

Rnd 33: Sl st in next 3 chs of next ch-7 lp; ch 3, dc in same ch as last sl st made; ch 3, sk next ch, 2 dc in next ch; * † ch 5, sc in 4th ch of next ch-7 lp, ch 7, sc in 4th ch of next ch-7 lp, ch 5 †, 2 dc in 3rd ch of next ch-7 lp; ch 3, sk next ch, 2 dc in next ch; rep from * 28 times more, then rep from † to † once; join in 3rd ch of beg ch-3.

Rnd 34: Sl st in next dc and in next ch-3 lp; ch 3, 8 dc in same lp: beg candle base made; * ch 7, sc in 4th ch of next ch-7 lp, picot; ch 7, 9 dc in next ch-3 lp: candle base made; rep from * 29 times more, ending last rep without working last 9 dc; join in 3rd ch of beg ch-3: 30 candle bases.

Rnd 35: Ch 3, dc in same ch as joining; * dc in next dc, ch 3, sk next 2 dc, sc in next dc, picot; ch 3, sk next 2 dc, dc in next dc, 2 dc in next dc; ch 3, sc in 4th ch of next ch-7 lp, ch 7, sc in 4th ch of next ch-7 lp, ch 3, 2 dc in next dc; rep from * 29 times more, ending last rep without working last 2 dc; join in 3rd ch of beg ch-3.

Rnd 36: Ch 3, dc in next 2 dc, * ch 2, sc in 2nd ch of next ch-3 lp, ch 7, sc in 2nd ch of next ch-3 lp, ch 2, dc in next 3 dc, ch 5, sc in 4th ch of next ch-7 lp, picot; ch 5, dc in next 3 dc; rep from * 29 times more, ending last rep without working last 3 dc; join in 3rd ch of beg ch-3.

Rnd 37: Ch 3, dc in next 2 dc, * ch 2, sk next ch of next ch-7 lp, sc in next 2 chs, 3 hdc in next ch; sc in next 2 chs, ch 2, dc in next 3 dc, ch 3, sc in 3rd ch of next ch-5 lp, ch 7, sc in 3rd ch of next ch-5 lp, ch 3, dc in next 3 dc; rep from * 29 times more, ending last rep without working last 3 dc; join in 3rd ch of beg ch-3.

Rnd 38: Ch 3, dc in next 2 dc, * ch 2, sk next 2 sc and next hdc; sc in next hdc, picot; ch 2, dc in next 3 dc, ch 7, sc in 4th ch of next ch-7 lp, picot; ch 7, dc in next 3 dc; rep from * 29 times more, ending last rep without working last 3 dc; join in 3rd ch of beg ch-3.

Rnd 39: Ch 3, dc in next 2 dc, * ch 3, dc in next 3 dc, ch 5, sc in 4th ch of next ch-7 lp, ch 7, sc in 4th ch of next ch-7 lp, ch 5, dc in next 3 dc; rep from * 29 times more, ending last rep without working last 3 dc; join in 3rd ch of beg ch-3.

Rnd 40: Ch 3, dc in next 2 dc, * ch 1, dc in next 3 dc, ch 3, sc in 3rd ch of next ch-5 lp, ch 7, sc in 4th ch of next ch-7 lp, picot; ch 7, sc in 3rd ch of next ch-5 lp, ch 3, dc in next 3 dc; rep from * 29 times more, ending last rep without working last 3 dc; join in 3rd ch of beg ch-3.

Rnd 41: Ch 3, * CL; sk next ch-1 sp, CL; dc in next dc, ch 7, (sc in 4th ch of next ch-7 lp, ch 7) twice; dc in next dc; rep from * 29 times more, ending last rep without working last dc; join in 3rd ch of beg ch-3.

Rnd 42: Ch 2, 3-dc CL; picot; * (ch 7, sc in 4th ch of next ch-7 lp) twice; picot in last sc made; ch 7, sc in 4th ch of next ch-7 lp, ch 7, 4-dc CL; picot; rep from * 29 times more, ending last rep without working last 4-dc CL and last picot; join in top of first 3-dc CL.

Rnd 43: Sl st in next 4 chs of next ch-7 lp; ch 1, sc in same ch as last sl st made; * ch 7, sl st in 4th ch from hook: picot made; ch 4, sc in 4th ch of next ch-7 lp; rep from * around, ending last rep without working last sc; join in first sc.

Finish off and weave in all ends.

Victorian Lace

designed by Jo Ann Maxwell

Size:

About 23" diameter

Materials:

Bedspread-weight crochet cotton, two 225-yd balls white; one 225-yd ball ecru

Size 7 steel crochet hook, or size required for gauge

Gauge:

9 dc = 1"

Pattern Stitch

Double Triple Crochet (dtrc)**:** YO 3 times, draw up lp in next st, (YO and draw through 2 lps on hook) 4 times: dtrc made.

Instructions

With white, ch 4; join to form a ring.

Rnd 1: Ch 3 (counts as a dc on this and following rnds), 15 dc in ring; join in 3rd ch of beg ch-3: 16 dc.

Rnd 2: Ch 8 (counts as a dtrc and a ch-3 lp); * dtrc (see Pattern Stitch) in next dc, ch 3; rep from * 14 times more; join in 4th ch of beg ch-8: 16 dtrc.

Rnd 3: Ch 3, dc in each ch and in each dtrc; join in 3rd ch of beg ch-3: 64 dc.

Rnd 4: Ch 1, sc in same ch as joining; ch 4, sl st in sc just made: picot made; ch 5, sk next 3 dc, * sc in next dc; ch 4, sl st in st just made: picot made; ch 5, sk next 3 dc; rep from * around; join in first sc: 16 ch-5 lps. Finish off.

Rnd 5: Join ecru in 3rd ch of next ch-5 lp; ch 3, 4 dc in same ch as joining; ch 3, * 5 dc in 3rd ch of next ch-5 lp; ch 3; rep from * around; join in 3rd ch of beg ch-3. Finish off.

Rnd 6: Join white in 2nd ch of next ch-3 lp; ch 1, sc in same lp; ch 7, * sc in next ch-3 lp, ch 7; rep from * around; join in first sc.

Rnd 7: Sl st in next 2 chs of next ch-7 lp; ch 3, dc in same ch as last sl st made; * 2 dc in next ch; 3 dc in next ch; 2 dc in each of next 2 chs; 2 dc in 2nd ch of next ch-7 lp; rep from * around, ending last rep without working last 2 dc; join in 3rd ch of beg ch-3.

Rnd 8: Sl st in next 5 dc; ch 1, sc in same dc as last sl st made; * ch 9, sk next 10 dc, sc in next dc; rep from * around, ending last rep without working last sc; join in first sc: 16 ch-9 lps. Finish off.

Rnd 9: Join ecru in 3rd ch of next ch-9 lp; ch 3, 4 dc in same ch as joining; ch 3, * sk next 3 chs, 5 dc in next ch; 5 dc in 3rd ch of next ch-9 lp; ch 3; rep from * around, ending last rep without working last 5 dc and ch 3; join in 3rd ch of beg ch-3. Finish off.

Rnd 10: Join white in any ch-3 lp; ch 6 (counts as a dc and a ch-3 lp on this and following rnds), dc in same lp as joining; * ch 4, sc between next two 5-dc groups, ch 4, in next ch-3 lp work (dc, ch 3, dc); rep from * around, ending last rep without working last (dc, ch 3, dc); join in 3rd ch of beg ch-6.

Rnd 11: Ch 1, sc in same ch as joining and in next 2 chs; ch 12, sl st in last sc made: ch-12 lp made; * sc in next ch, in next dc, in next 4 chs, in next sc, in next 4 chs, in next dc, and in next 2 chs; ch 12, sl st in last sc made: ch-12 lp made; rep from * 16 times more, ending last rep without working last 3 sc and last ch-12 lp; join in first sc: 16 ch-12 lps.

Rnd 12: Sl st in next sc; * working around next ch-12 lp, 3 dc in next ch; 2 dc in each of next 4 chs; 3 dc in each of next 2 chs; 2 dc in each of next 4 chs; 3 dc in next ch; sk next sc, sl st in next 6 sc; picot; sl st in next 7 sc; rep from * around, ending last rep without working last sl st; join in bl of first sl st.

Rnd 13: Working in bls only, sl st in next sl st and in next 10 dc; ch 5 (counts as a dc and a ch-2 sp), * sk next dc, dc in next dc, ch 2, sk next dc, dc in next dc, ch 3, dc in next dc, ch 2, sk next dc, dc in next dc, ch 2, sk next dc, dc in next dc, ch 3, dc in 10th dc of next ch-12 lp, ch 2; rep from * around, ending last rep without working last dc and ch 2; join in 3rd ch of beg ch-5.

Rnd 14: (Sl st in next 2 chs and in next dc) twice; sl st in next 2 chs of next ch-3 lp; ch 1, sc in same ch as last sl st made; picot; ch 5, in 2nd ch of next ch-3 lp work (trc, picot, ch 4, trc, picot); ch 5, * sc in 2nd ch of next ch-3 lp, picot; ch 5, in 3rd ch of next ch-3 lp work (trc, picot, ch 4, trc, picot); ch 5; rep from * around; join in first sc.

Rnd 15: Sl st in next 3 chs of next ch-5 lp and in same lp; ch 8 (counts as a dc and a ch-5 lp); * dc in next lp, ch 5; rep from * around; join in 3rd ch of beg ch-8.

Rnd 16: Ch 3, dc in each st and in each ch; join in 3rd ch of beg ch-3: 216 dc.

continued on page 26

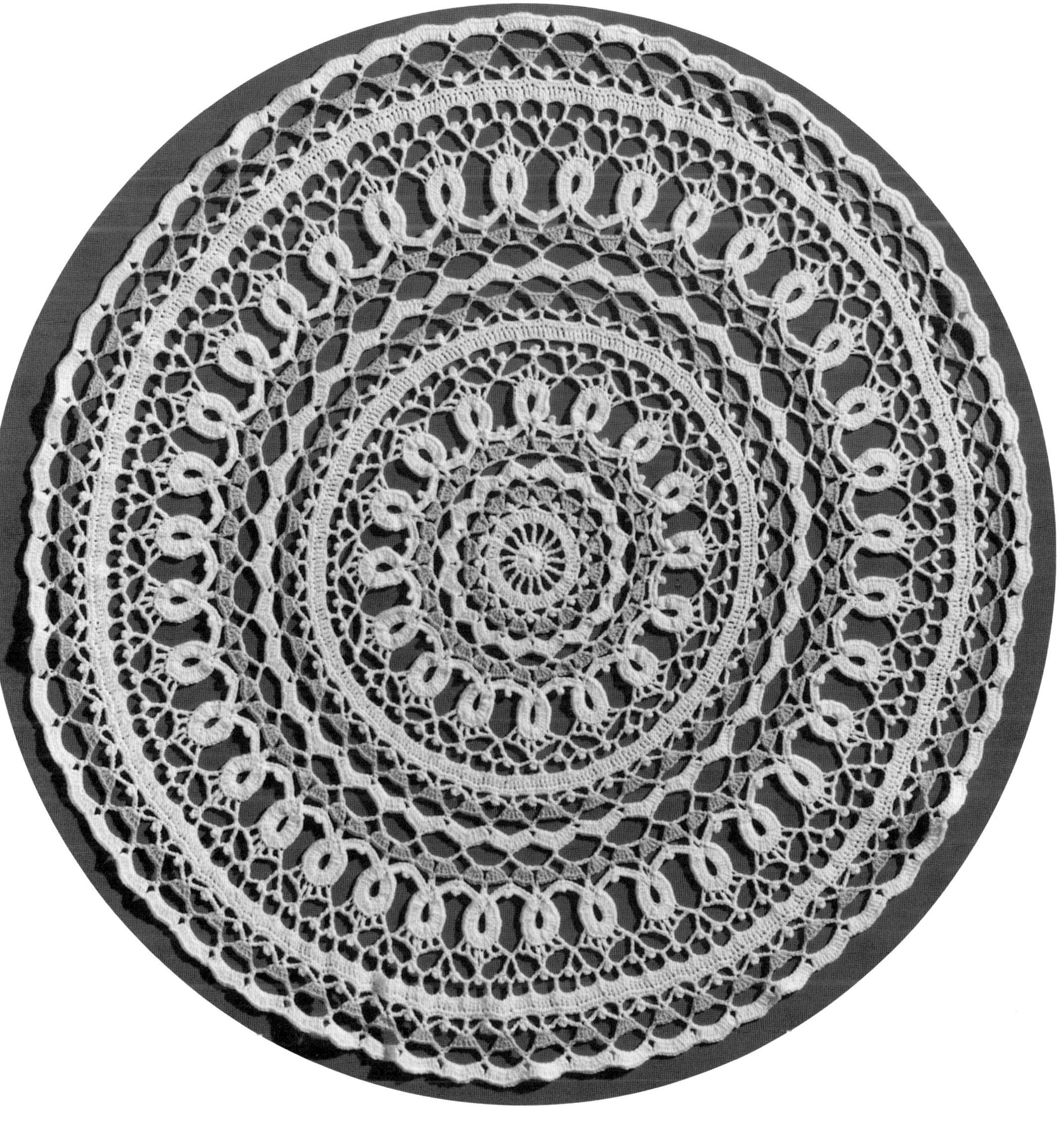

Rnd 17: Ch 1, sc in same ch as joining; sc in each dc; join in first sc.

Rnd 18: Ch 7 (counts as a dc and a ch-4 lp); * sk next 3 sc, sc in next sc, picot; ch 4, sk next 3 sc, dc in next sc, ch 4; rep from * around, ending last rep without working last dc and ch 4; join in 3rd ch of beg ch-7. Change to ecru by drawing lp through; cut white.

Rnd 19: Ch 3, 4 dc in same ch as joining; ch 5, * 5 dc in next dc; ch 5; rep from * around; join in 3rd ch of beg ch-3. Finish off.

Rnd 20: Join white in 3rd ch of next ch-5 lp; ch 1, sc in same ch as joining; ch 9, * sc in 3rd ch of next ch-5 lp, ch 9; rep from * around; join in first sc.

Rnd 21: Sl st in next 2 chs of next ch-9 lp; ch 3, dc in same ch as last sl st made; * dc in next ch, 2 dc in next ch; 3 dc in next ch; 2 dc in next ch; dc in next ch; 2 dc in next ch; 2 dc in 2nd ch of next ch-9 lp; rep from * around, ending last rep without working last 2 dc; join in 3rd ch of beg ch-3.

Rnd 22: Sl st in next 6 dc; ch 1, sc in same dc as last sl st made; ch 11, sk next 12 dc, * sc in next dc, ch 11, sk next 12 dc; rep from * around; join in first sc. Finish off.

Rnd 23: Join ecru in 4th ch of next ch-11 lp; ch 3, 4 dc in same ch as joining; * ch 3, sk next 3 chs, 5 dc in next ch; ch 1, 5 dc in 4th ch of next ch-11 lp; rep from * around, ending last rep without working last 5 dc; join in 3rd ch of beg ch-3. Finish off.

Rnd 24: Join white in 2nd ch of next ch-3 lp; ch 6, dc in same ch as joining; ch 5, sc in next ch-1 sp, ch 5, * in 2nd ch of next ch-3 lp work (dc, ch 3, dc); ch 5, sc in next ch-1 sp, ch 5; rep from * around; join in 3rd ch of beg ch-6.

Rnd 25: Ch 1, sc in same ch as joining and in next 2 chs; ch 12, sl st in last sc made: ch-12 lp made; sc in next ch and in next dc; * sc in next 14 sts, ch 12, sl st in last sc made: ch-12 lp made; sc in next ch and in next dc; rep from * 34 times more; sc in next 11 sts; join in first sc.

Rnd 26: Sl st in next ch of next ch-12 lp; ch 3, 2 dc in same ch; * working around same ch-12 lp, 2 dc in each of next 4 chs; 3 dc in each of next 2 chs; 2 dc in each of next 4 chs; 3 dc in next ch; sk next sc, sl st in next 6 sc; sl st in next sc, picot; sl st in next 6 sc; 3 dc in first ch of next ch-12 lp; rep from * around, ending last rep without working last 3 dc; join in 3rd ch of beg ch-3.

Rnds 27 through 29: Rep Rnds 13 through 15.

Rnd 30: Ch 2, dc in next 5 chs, dec over next dc and next ch [to work dec: (YO, draw up lp in next st, YO and draw through 2 lps on hook) twice; YO and draw through all 3 lps on hook: dec made]; * dc in next 4 chs, dec as before; rep from * around; join in first dc: 540 dc.

Rnd 31: Ch 1, sc in same dc as joining; * sc dec over next 2 dc (to work sc dec: draw up lp in each of next 2 dc, YO and draw through all 3 lps on hook: sc dec made); sc in next 133 sc; rep from * 3 times more, ending last rep without working last sc; join in first sc.

Rnds 32 through 35: Rep Rnds 18 through 21.

Finish off and weave in all ends.

Gentle Wind

designed by Jo Ann Maxwell

Size:
About 26" diameter

Materials:
Bedspread-weight crochet cotton, two 225-yd balls white; 100 yds ecru
Size 6 steel crochet hook, or size required for gauge

Gauge:
8 dc = 1"

Pattern Stitches
Double Triple Crochet (dtrc)**:** YO 3 times, draw up lp in next st, (YO and draw through 2 lps on hook) 4 times: dtrc made.
Lover's Knot (LK)**:** Draw up lp on hook 1/2", YO and draw through beg 1/2" lp just made, insert hook in lp on left side (not through beg 1/2" lp) as in **Fig 1**, YO and draw through, YO and draw through 2 lps on hook: LK made.

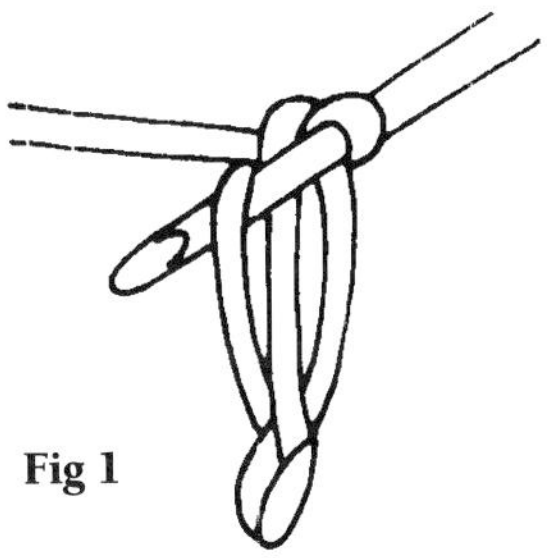

Fig 1

Instructions
With white, ch 5; join to form a ring.

Rnd 1: Ch 3 (counts as a dc on this and following rnds), 19 dc in ring; join in 3rd ch of beg ch-3: 20 dc.

Rnd 2: Ch 5 (counts as a dc and a ch-2 sp), * dc in next dc, ch 2; rep from * 18 times more; join in 3rd ch of beg ch-5.

Rnd 3: Ch 3, dc in each ch and in each dc; join in 3rd ch of beg ch-3: 60 dc.

Rnd 4: Ch 2 (counts as an hdc on this and following rnds), hdc in same ch as joining; ch 3, sk next 2 dc, * 2 hdc in next dc; ch 3, sk next 2 dc; rep from * around; join in 2nd ch of beg ch-2: 20 ch-3 lps. Finish off.

Rnd 5: Join ecru in any ch-3 lp; ch 1, sc in same lp; LK (see Pattern Stitches); * sc in next ch-3 lp, ch 1, LK; rep from * around; join in first sc by drawing white through; cut ecru.

Rnd 6: Ch 1, sc in same sc as joining; ch 5, * sc in next sc, ch 5; rep from * around; join in first sc.

Rnd 7: * Sc in next ch, hdc in next ch, 3 hdc in next ch; hdc in next ch, sc in next ch, sl st in next sc; rep from * around, ending last rep without working last sl st; join in bl of joining sl st of Rnd 6.

Rnd 8: Working in bls only, sl st in next sc and in next 3 hdc; ch 2, hdc in same hdc as last sl st made; ch 5, * sk next 7 sts, 2 hdc in next hdc; ch 5; rep from * around; join in 2nd ch of beg ch-2.

Rnd 9: Sl st in next hdc, in next 3 chs of next ch-5 lp and in same lp; ch 1, sc in same lp; ch 3, sl st in sc just made: picot made; ch 7, * sc in next ch-5 lp; ch 3, sl st in sc just made: picot made; ch 7; rep from * around; join in first sc.

Rnd 10: Sl st in next 4 chs of next ch-7 lp and in same lp; ch 1, sc in same lp; ch 7, * sc in next ch-7 lp, ch 7; rep from * around; join in first sc.

Rnd 11: Sl st in next ch; ch 3, * dc in next ch, 2 dc in next ch; 3 dc in next ch; 2 dc in next ch; dc in next 2 chs; dc in next ch of next ch-7 lp; rep from * around, ending last rep without working last dc; join in 3rd ch of beg ch-3.

Rnd 12: Sl st in bls of next 2 dc; ch 2, working in bls only, hdc in same dc as last sl st made; (ch 3, sk next 2 dc, 2 hdc in next dc) twice; * ch 3, sk next 4 dc, 2 hdc in next dc; (ch 3, sk next 2 dc, 2 hdc in next dc) twice; rep from * around; ch 3; join in 2nd ch of beg ch-2.

Rnd 13: Sl st in next hdc and in next 2 chs of next ch-3 lp; ch 2, hdc in same ch as last sl st made; ch 3, * 2 hdc in 2nd ch of next ch-3 lp; ch 3; rep from * around; join in 2nd ch of beg ch-2.

Rnds 14 through 16: Rep Rnd 13. At end of Rnd 16, finish off.

Rnds 17 through 20: Rep Rnds 5 through 8.

Rnd 21: Sl st in next hdc and in first ch of next ch-5 lp; ch 6 (counts as a trc and a ch-2 sp), trc in next ch, (ch 2, trc in next ch) 3 times; sc in next ch-5 lp, ch 5, sc in next ch-5 lp, * trc in first ch of next ch-5 lp, (ch 2, trc in next ch) 4 times; sc in next ch-5 lp, ch 5, sc in next ch-5 lp; rep from * around; join in 4th ch of beg ch-6.

continued on page 29

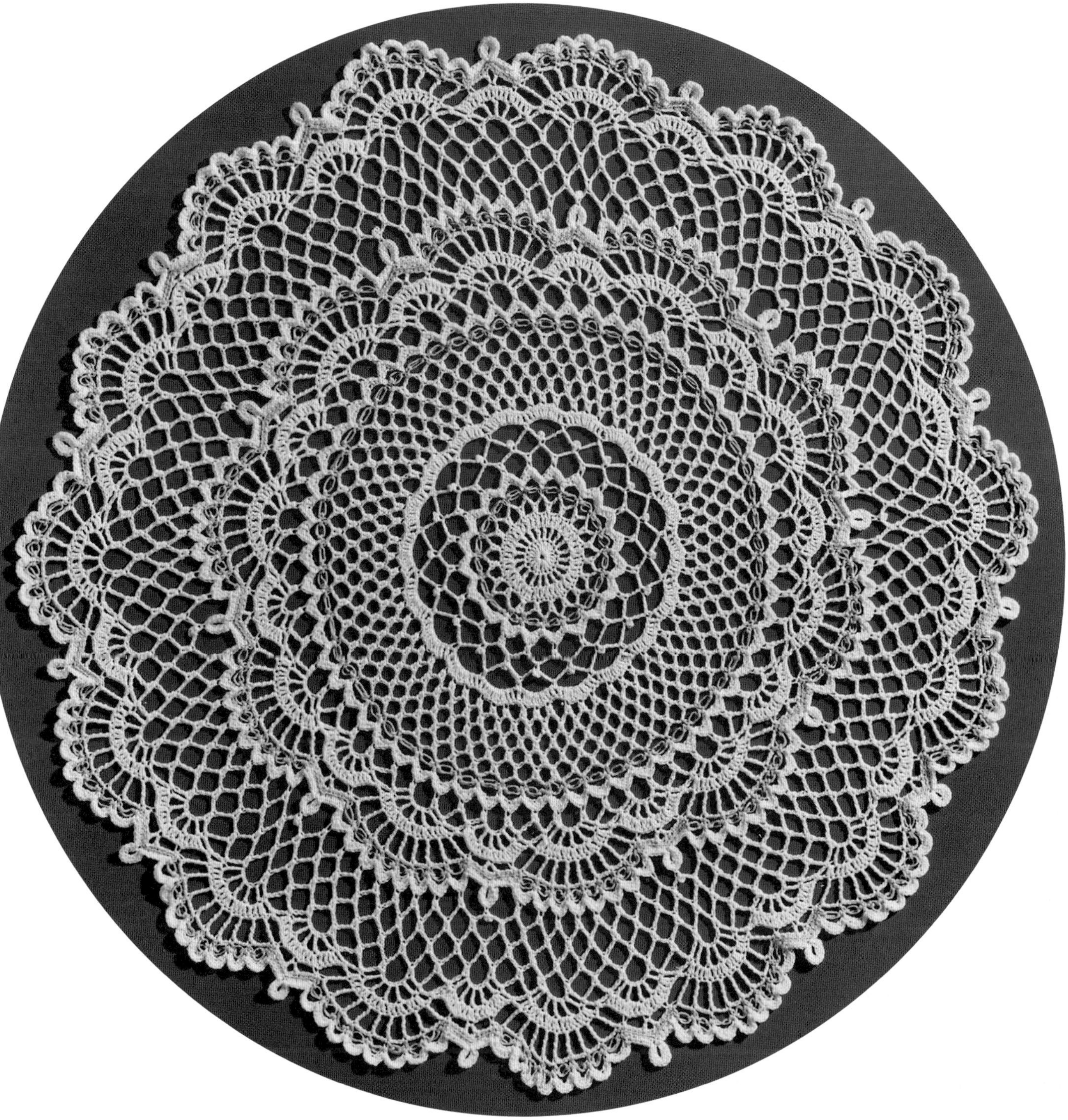

Rnd 22: Ch 3, dc in same ch as joining; * (dc in next 2 chs, 2 dc in next trc) 4 times; ch 1, sc in next ch-5 lp, ch 1, 2 dc in next trc; rep from * around; join in 3rd ch of beg ch-3.

Rnd 23: Sl st in next dc; ch 8 (counts as a dtrc and a ch-3 lp), sk next dc, * [dtrc (see Pattern Stitches, page 27) in next dc, ch 3, sk next dc] 7 times; dtrc in next dc, sk next sc and next 5 dc; sc in bl of next dc, ch 3, sk next 3 dc, in bl of next dc work (dc, ch 3, dc); ch 3, sk next 3 dc, sc in bl of next dc, sk next 4 dc, next sc, and next dc; dtrc in next dc, ch 3, sk next dc; rep from * around, ending last rep without working last dtrc and ch 3; join in 5th ch of beg ch-8 by drawing ecru through; cut white.

Rnd 24: Ch 1, sc in same ch as joining; * (ch 1, LK, sc in next dtrc) 8 times; sc in next sc, (sc in next 3 chs and in next dc) twice; sc in next 3 chs, in next sc, and in next dtrc; rep from * around, ending last rep without working last sc; join in first sc by drawing white through; cut ecru.

Rnd 25: Ch 1, sc in same sc as joining; * (ch 5, sk next LK, sc in next sc) 8 times; sc in bls of next 7 sc, ch 6, sl st in last sc made; sc in bls of next 7 sts; rep from * around, ending last rep without working last sc; join in first sc.

Rnd 26: Sl st in first ch of next ch-5 lp; ch 1, sc in same ch; hdc in next ch, 3 hdc in next ch; hdc in next ch, sc in next ch, sl st in next sc; * † sc in first ch of next ch-5 lp, hdc in next ch, 3 hdc in next ch; hdc in next ch, sc in next ch, sl st in next sc †; rep from † to † 6 times more; sk next 2 sc, sc in next 4 sc, sl st in next sl st; working around next ch-6 lp, 2 sc in each of next 6 chs; sl st in next sl st; sc in next 4 sc, sk next 2 sc, sl st in next sc; rep from * around, ending last rep without working last sl st; join in first sc.

Rnd 27: Sl st in bl of next sc and in bls of next 3 hdc; ch 2, hdc in same lp as last sl st made; * (ch 5, sk next 7 sts, 2 hdc in bl of next hdc) 7 times; ch 5, sc in bl of 6th sc of next ch-6 lp, picot; ch 5, 2 hdc in bl of 3rd hdc of next 5-hdc group; rep from * around, ending last rep without working last 2 hdc; join in 2nd ch of beg ch-2.

Rnd 28: Sl st in next hdc and in next 3 chs of next ch-5 lp; ch 2, hdc in same ch as last sl st made; ch 5, * 2 hdc in 3rd ch of next ch-5 lp; ch 5; rep from * around; join in 2nd ch of beg ch-2.

Rnds 29 and 30: Rep Rnd 28.

Rnds 31 through 36: Rep Rnds 21 through 26. At end of Rnd 36, finish off and weave in all ends.

Antique Lace

designed by Kathy Wigington

Size:

About 13" diameter

Materials:

Bedspread-weight crochet cotton, two 225-yd balls ecru; 50 yds each rose and green
Size 7 steel crochet hook, or size required for gauge
Size 18 tapestry needle

Gauge:

9 dc = 1"

Pattern Stitch

Double Triple Crochet (dtrc)**:** YO 3 times, draw up lp in next st, (YO and draw through 2 lps on hook) 4 times: dtrc made.

Instructions

Doily

With ecru, ch 26; join to form a ring.

Rnd 1: Ch 4, sk next ch, (sl st in next ch, ch 4, sk next ch) 12 times; join in first ch of beg ch-4: 13 ch-4 lps.

Rnd 2: Sl st in next 2 chs of next ch-4 lp and in same lp; ch 4, (sl st in next ch-4 lp, ch 4) 12 times; join in 3rd sl st.

Rnd 3: Sl st in next 2 chs of next ch-4 lp and in same lp; ch 5, (sl st in next ch-4 lp, ch 5) 12 times; join in 3rd sl st.

Rnd 4: Sl st in next 3 chs of next ch-5 lp and in same lp; ch 6, (sl st in next ch-5 lp, ch 6) 12 times; join in 4th sl st.

Rnd 5: Sl st in next 3 chs of next ch-6 lp and in same lp; ch 7, (sl st in next ch-6 lp, ch 7) 12 times; join in 4th sl st.

Rnd 6: Sl st in next 4 chs of next ch-7 lp and in same lp; ch 8, (sl st in next ch-7 lp, ch 8) 12 times; join in 5th sl st.

Rnd 7: Sl st in next 4 chs of next ch-8 lp and in same lp; ch 9, (sl st in next ch-8 lp, ch 9) 12 times; join in 5th sl st.

Rnd 8: Sl st in next 5 chs of next ch-9 lp and in same lp; ch 10, (sl st in next ch-9 lp, ch 10) 12 times; join in 6th sl st.

Rnd 9: Sl st in next ch-10 lp; ch 1, 10 sc in same lp; ch 1, (10 sc in next ch-10 lp, ch 1) 12 times; join in first sc: 130 sc.

Rnd 10: Ch 4, working in bls only, sk next sc, (sl st in next sc, ch 4, sk next sc) 64 times; join in joining sl st of prev rnd: 65 ch-4 lps.

Rnd 11: Sl st in next 2 chs of next ch-4 lp and in same lp; ch 4, (sl st in next ch-4 lp, ch 4) 64 times; join in 3rd sl st.

Rnds 12 and 13: Rep Rnd 11.

Rnd 14: Sl st in next 2 chs of next ch-4 lp and in same lp; ch 5, (sl st in next ch-4 lp, ch 5) 64 times; join in 3rd sl st.

Rnd 15: Sl st in next 3 chs of next ch-5 lp and in same lp; ch 5, (sl st in next ch-5 lp, ch 5) 64 times; join in 4th sl st.

Rnds 16 through 18: Rep Rnd 15.

Rnd 19: Sl st in next ch-5 lp; ch 1, 5 sc in same lp; 5 sc in each rem ch-5 lp; join in first sc: 325 sc.

Rnd 20: Ch 3 (counts as a dc on this and following rnds), 2 dc in same sc as joining; 3 dc in each rem sc; join in 3rd ch of beg ch-3.

Rnd 21: Ch 3, dc in each dc; join in 3rd ch of beg ch-3.

Rnd 22: Ch 4, * sk next dc, sl st in next dc; ch 4; rep from * around; join in joining sl st of prev rnd.

Rnd 23: Sl st in next 2 chs of next ch-4 lp and in same lp; ch 4, * sl st in next ch-4 lp; ch 4; rep from * around; join in 3rd sl st.

Rnd 24: Rep Rnd 23. Finish off.

Rnd 25: Join ecru in first unused lp of Rnd 9; ch 3, 2 dc in same lp as joining; working in rem unused lps, 3 dc in each lp; join in 3rd ch of beg ch-3.

Rnds 26 through 29: Rep Rnds 21 through 24.

Weave in all ends.

Rose

With rose, ch 2.

Rnd 1: 5 sc in 2nd ch from hook; join in first sc: 5 sc.

Rnd 2: Ch 1, working in fls only, in each sc work (sc, hdc, dc, hdc, sc): petal made; join in first sc: 5 petals.

Rnd 3: Ch 3, working behind petals made in prev rnd in unused lps of Rnd 1, * sl st in next lp; ch 3; rep from * 4 times more; join in joining sl st of Rnd 2.

Rnd 4: Sl st in next ch-3 lp; in same lp work (sc, hdc, 3 dc, hdc, sc): petal made; in each rem ch-3 lp work (sc, hdc, 3 dc, hdc, sc): petal made; join in first sc: 5 petals.

continued on page 32

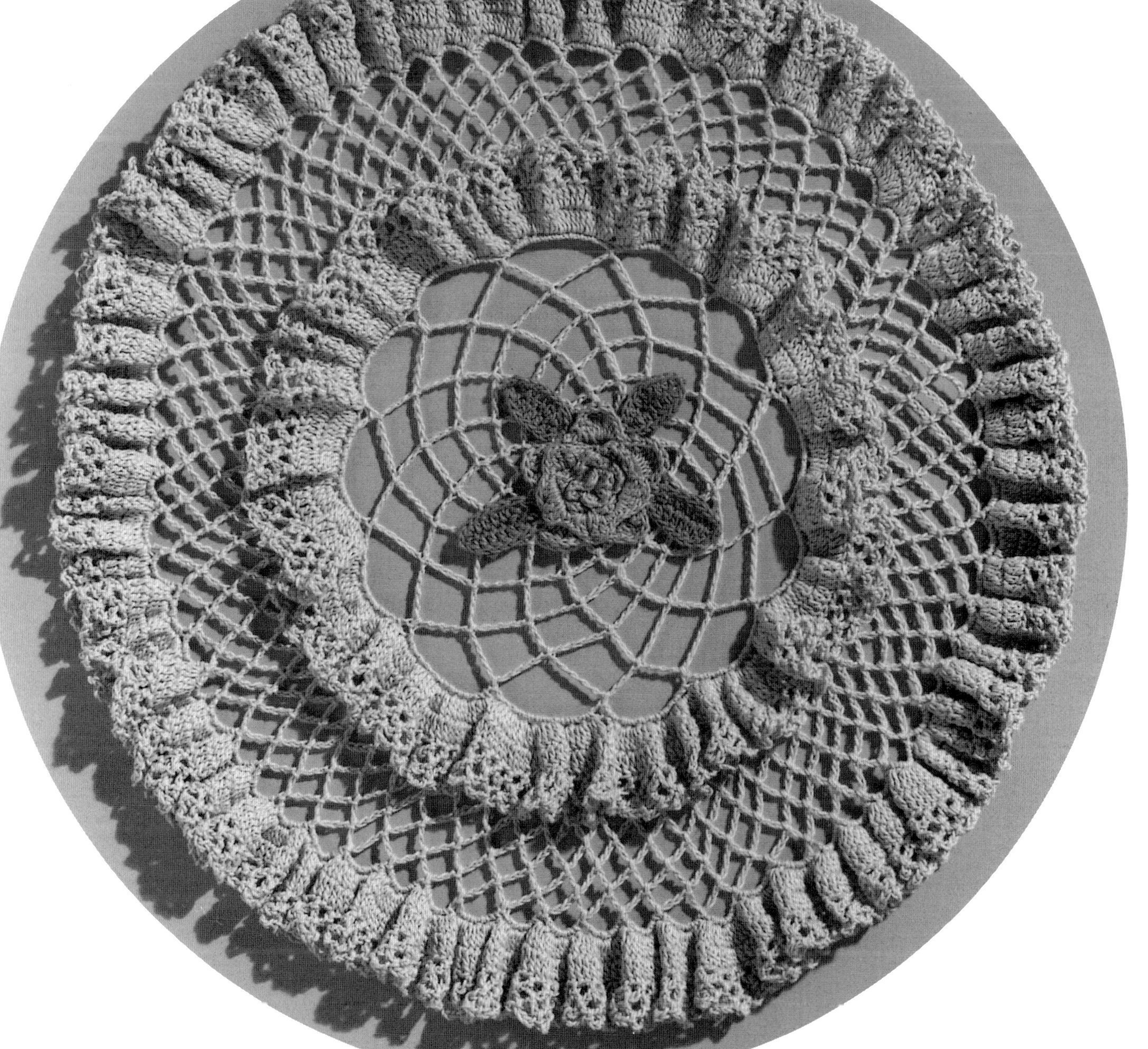

Rnd 5: Ch 2, working behind petals made in prev rnd, * sl st around post (see page 5) of 2nd dc of next petal; ch 4; rep from * 4 times more; join in first sl st: 5 ch-4 lps.

Rnd 6: Sl st in next ch-4 lp; in same lp work (sc, hdc, dc, 3 trc, dc, hdc, sc): petal made; in each rem ch-4 lp work (sc, hdc, dc, 3 trc, dc, hdc, sc): petal made; join in first sc: 5 petals.

Rnd 7: Ch 3, working behind petals made in prev rnd, * sl st around post of 2nd trc of next petal; ch 4; rep from * 4 times more; join in first sl st: 5 ch-4 lps.

Rnd 8: Sl st in next ch-4 lp, in same lp work [sc, hdc, dc, trc, 3 dtrc (see Pattern Stitch, page 30), trc, dc, hdc, sc]; in each rem ch-4 lp work (sc, hdc, dc, trc, 3 dtrc, trc, dc, hdc, sc); join in first sc: 5 petals. Finish off, leaving a 12" end for sewing.

Leaf (make 4)

With green, ch 9.

Rnd 1: Sc in 2nd ch from hook, hdc in next ch, dc in next 5 chs, 5 hdc in next ch; working in unused lps on opposite side of beg ch, dc in next 5 lps, hdc in next lp, sc in next ch; join in first sc. Finish off and weave in ends.

Finishing

With tapestry needle, sew 4 leaves to rose. Tack rose to center of doily.

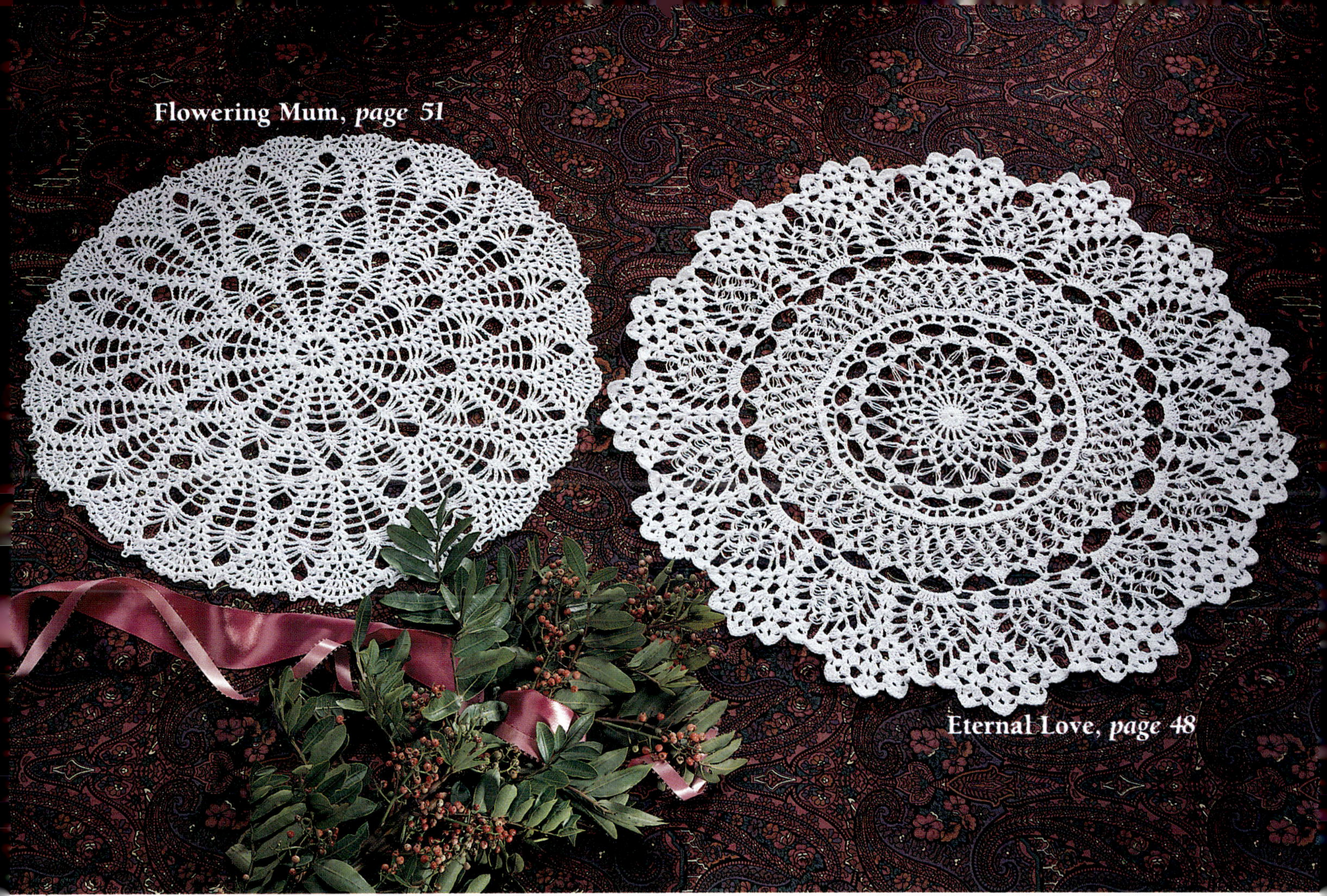

Flowering Mum, *page 51*

Eternal Love, *page 48*

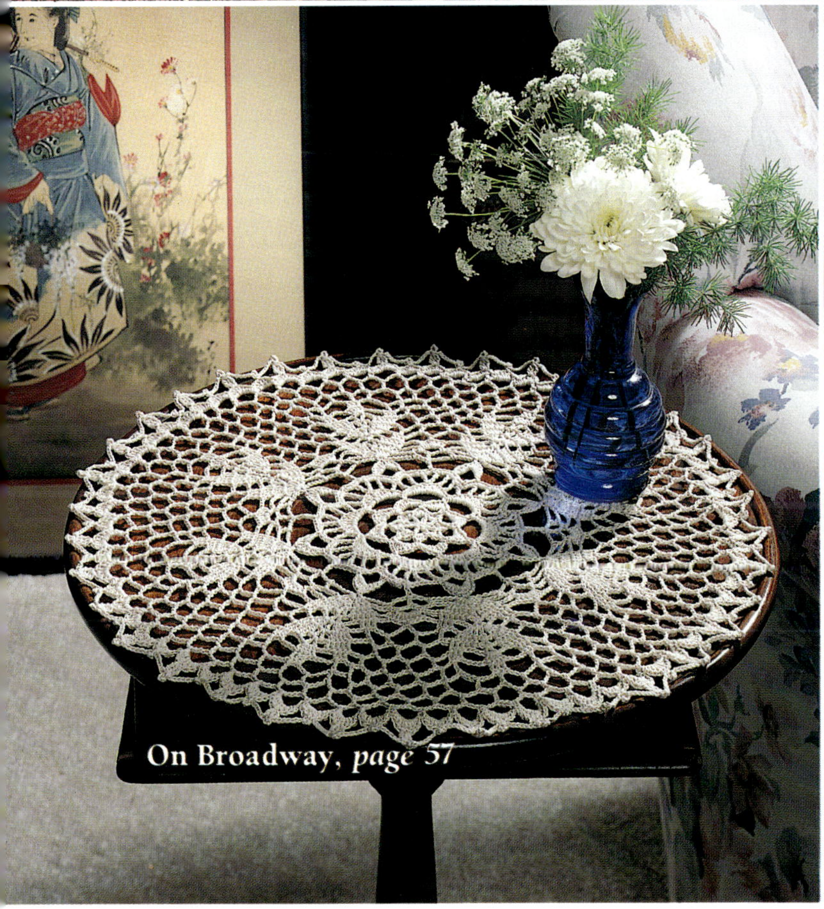

On Broadway, *page 57*

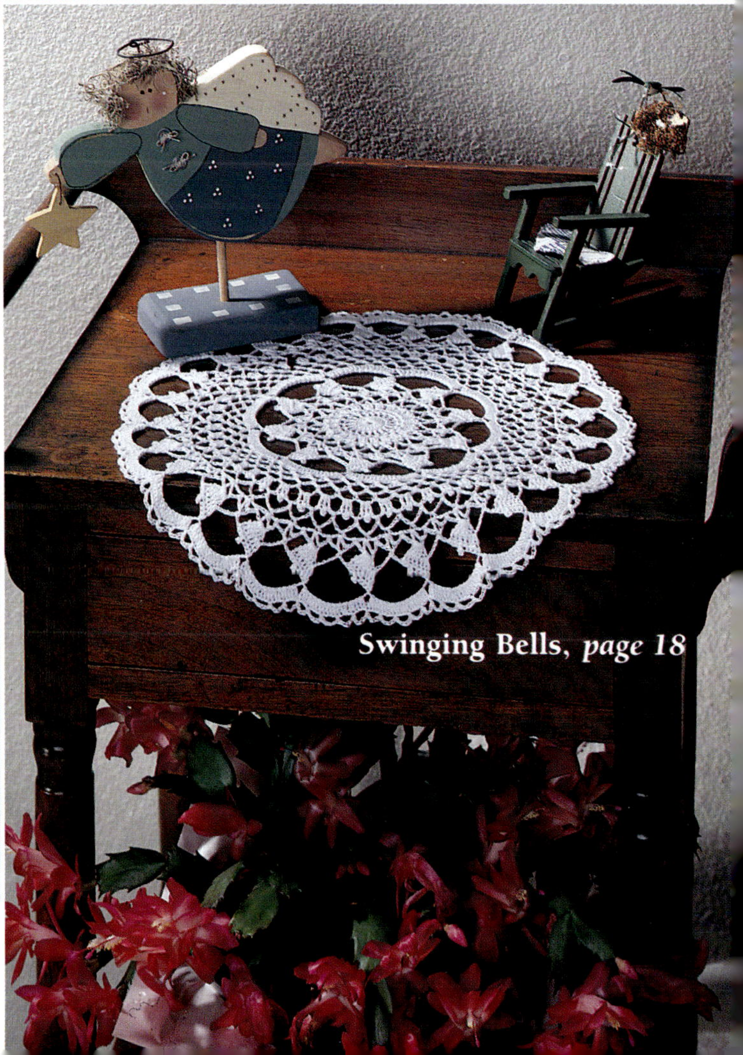

Swinging Bells, *page 18*

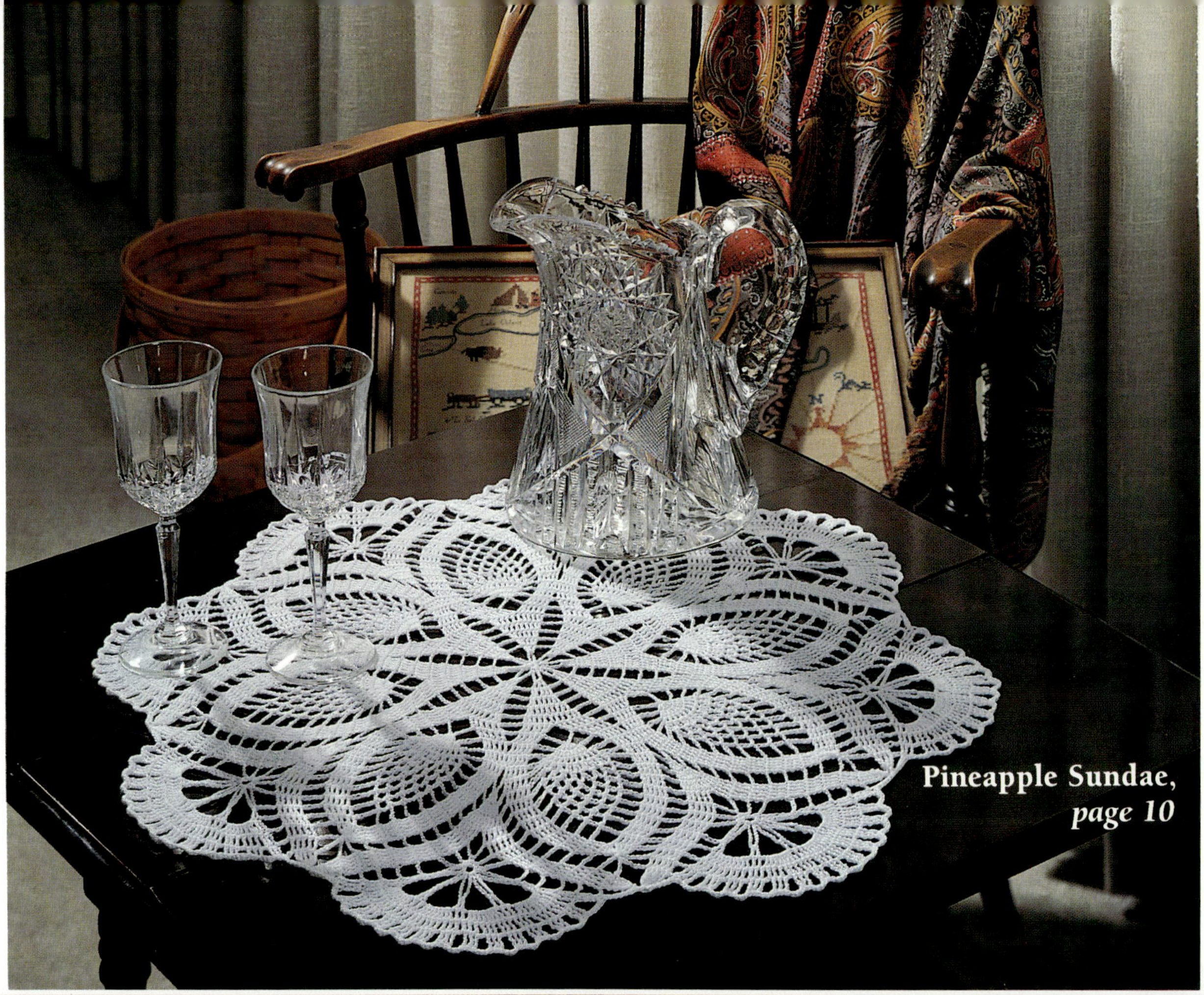

Pineapple Sundae, *page 10*

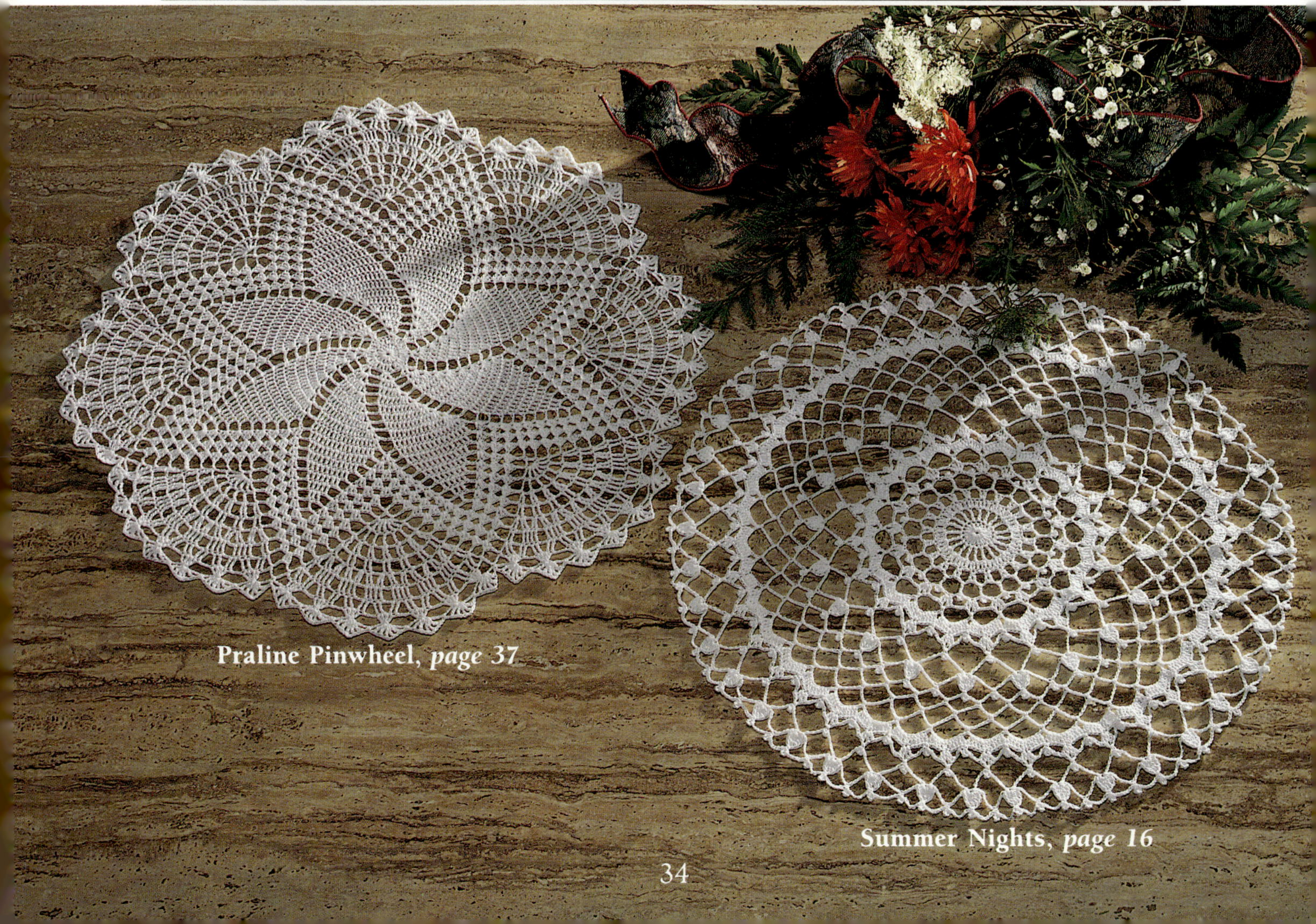

Praline Pinwheel, *page 37*

Summer Nights, *page 16*

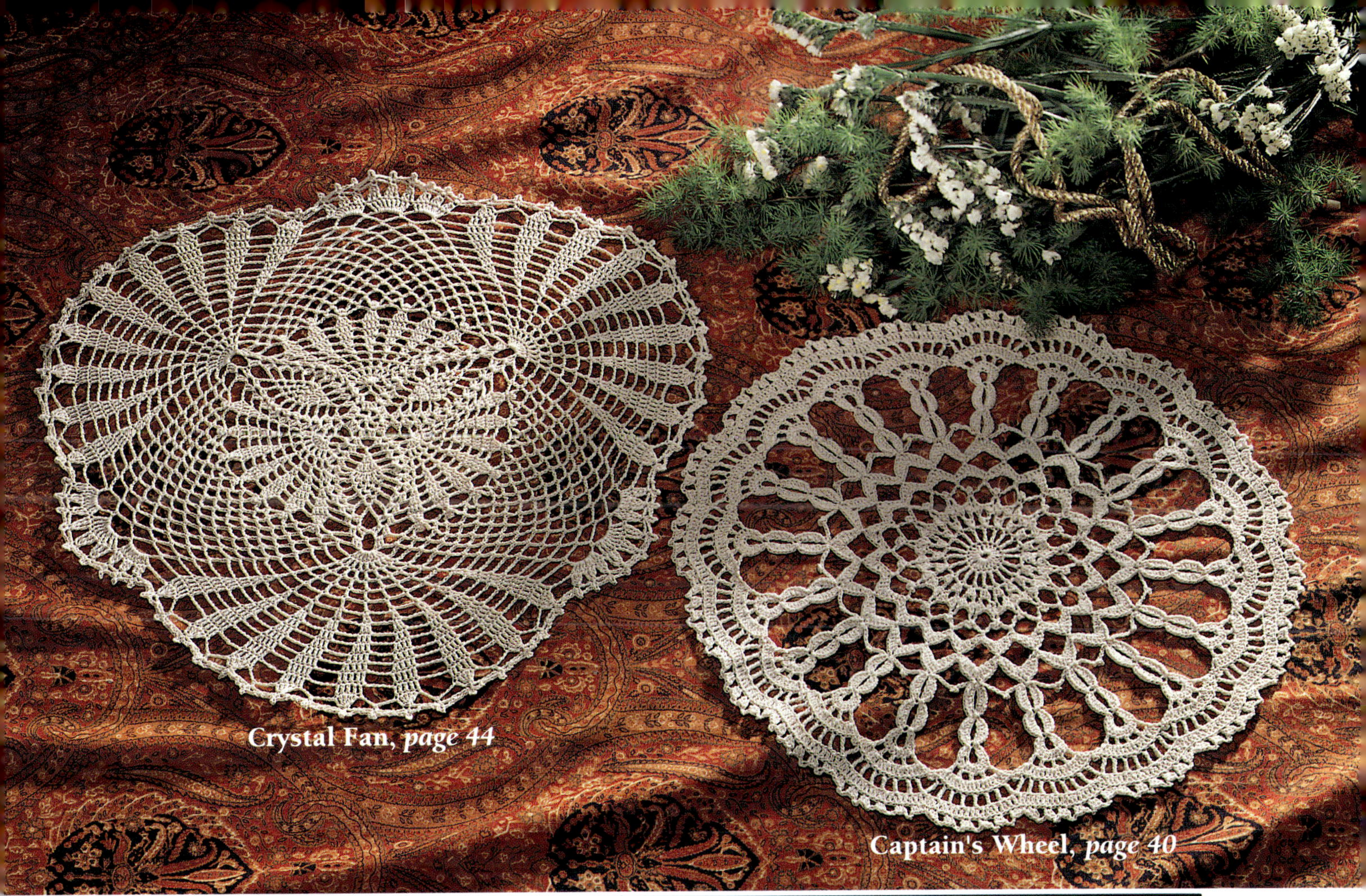

Crystal Fan, *page 44*

Captain's Wheel, *page 40*

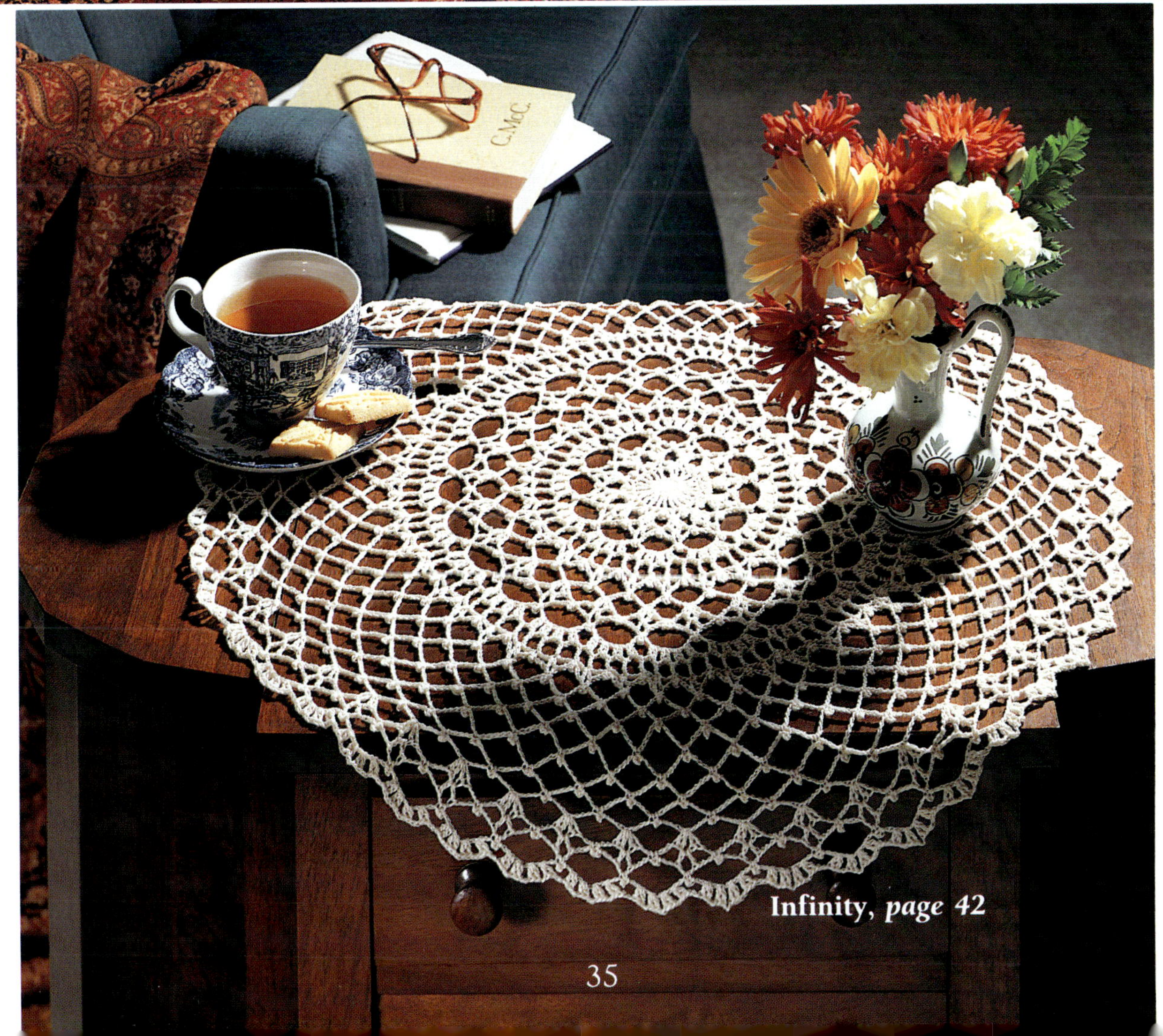

Infinity, *page 42*

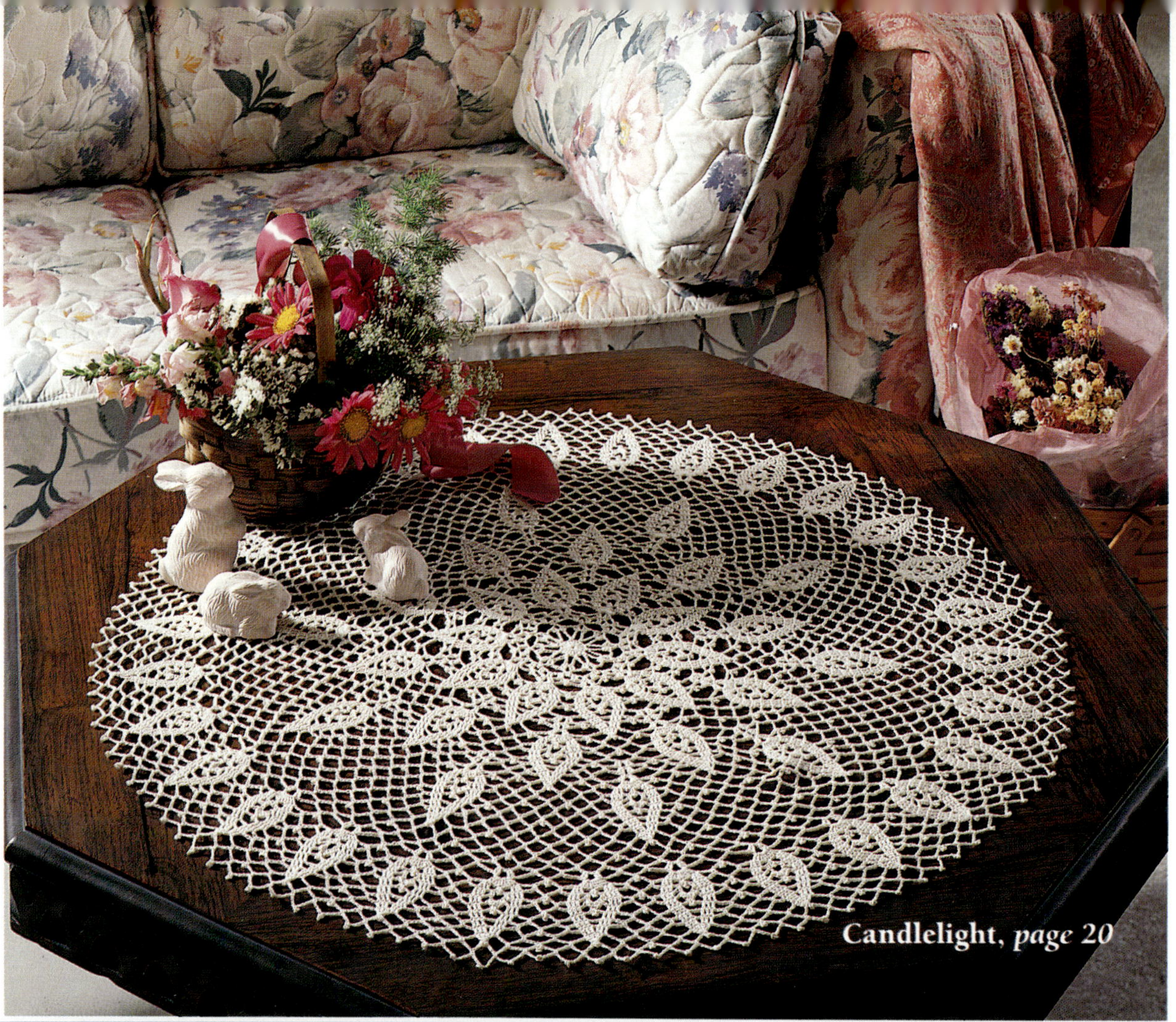

Candlelight, *page 20*

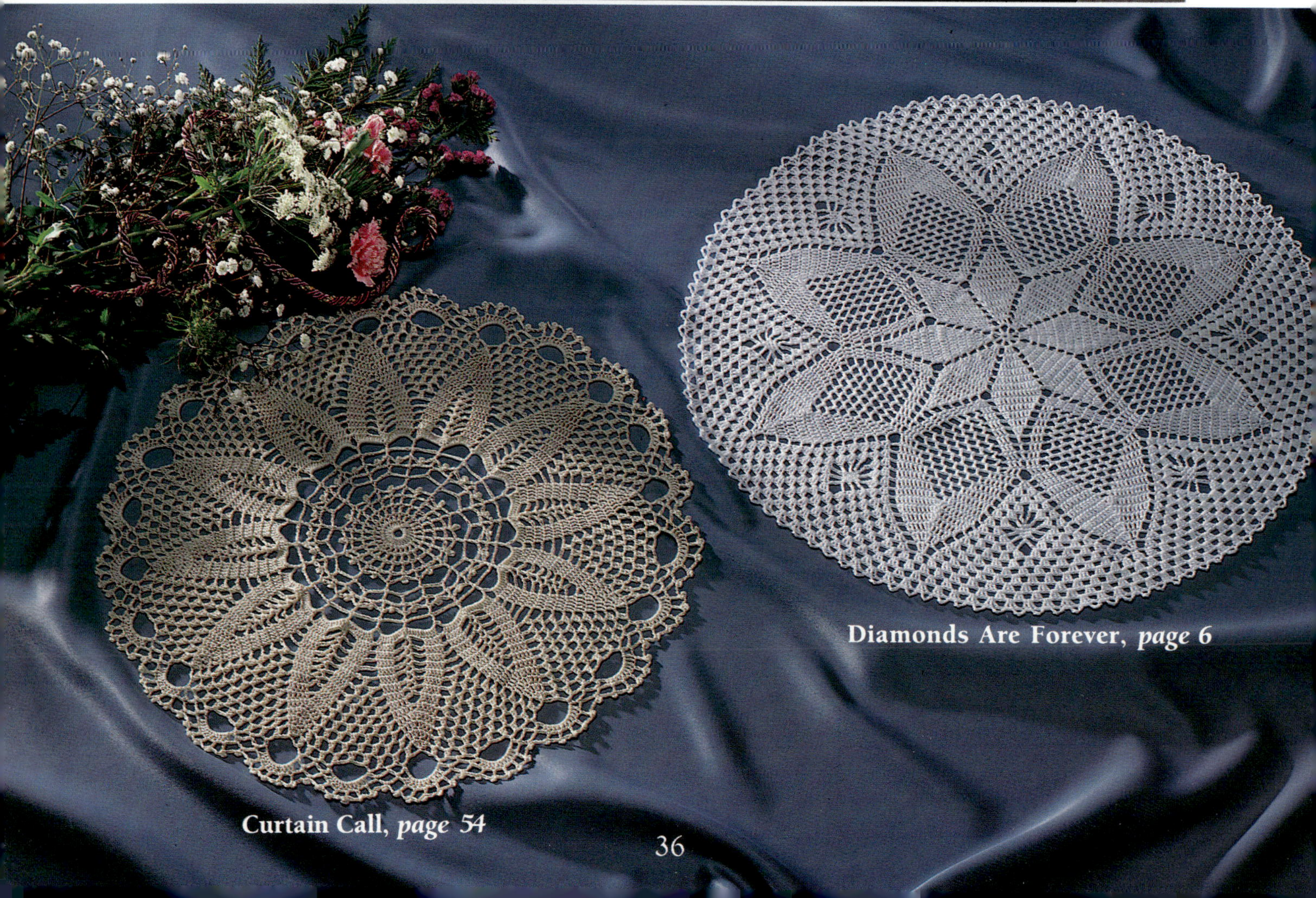

Diamonds Are Forever, *page 6*

Curtain Call, *page 54*

Praline Pinwheel

designed by Yalanda Wiese

Size:
About 18" diameter

Materials:
Bedspread-weight crochet cotton, two 225-yd balls white
Size 6 steel crochet hook, or size required for gauge

Gauge:
8 dc = 1"

Pattern Stitches
Beg Cluster (beg CL)**:** Ch 2, keeping last lp of each dc on hook, dc in next 3 trc, YO and draw through all 4 lps on hook: beg CL made.
Cluster (CL)**:** Keeping last lp of each dc on hook, dc in next 4 trc, YO and draw through all 5 lps on hook: CL made.

Instructions
Ch 6, join to form a ring.

Rnd 1: Ch 3, (counts as a dc on this and following rnds), 15 dc in ring; join in 3rd ch of beg ch-3: 16 dc.

Rnd 2: Ch 3, dc in same ch as joining; 2 dc in each dc; join in 3rd ch of beg ch-3: 32 dc.

Rnd 3: Ch 3, in same ch as joining work (dc, ch 1, dc): beg pinwheel base made; ch 2, * sk next 3 dc, in next dc work (2 dc, ch 1, dc): pinwheel base made; ch 2; rep from * 6 times more; join in 3rd ch of beg ch-3: 8 pinwheel bases.

Rnd 4: Sl st in next dc; ch 3, * in next ch-1 sp work (3 dc, ch 1, dc); ch 3, sk next ch-2 sp and next dc, dc in next dc; rep from * 7 times more, ending last rep without working last dc; join in 3rd ch of beg ch-3.

Rnd 5: Sl st in next dc; ch 3, dc in next 2 dc, in next ch-1 sp work (3 dc, ch 1, dc); ch 3, * sk next ch-3 lp and next dc, dc in next 3 dc, in next ch-1 sp work (3 dc, ch 1, dc); ch 3; rep from * 6 times more; join in 3rd ch of beg ch 3.

Rnd 6: Sl st in next dc; ch 3, dc in next 4 dc, in next ch-1 sp work (3 dc, ch 1, dc); ch 3, * sk next ch-3 lp and next dc, dc in next 5 dc, in next ch-1 sp work (3 dc, ch 1, dc); ch 3; rep from * 6 times more; join in 3rd ch of beg ch-3.

Rnd 7: Sl st in next dc; ch 3, dc in next 6 dc, in next ch-1 sp work (3 dc, ch 1, dc); ch 3, * sk next ch-3 lp and next dc, dc in next 7 dc, in next ch-1 sp work (3 dc, ch 1, dc); ch 3; rep from * 6 times more; join in 3rd ch of beg ch-3.

Rnd 8: Sl st in next dc; ch 3, dc in next 8 dc, in next ch-1 sp work (3 dc, ch 1, dc); ch 3, * sk next ch-3 lp and next dc, dc in next 9 dc, in next ch-1 sp work (3 dc, ch 1, dc); ch 3; rep from * 6 times more; join in 3rd ch of beg ch-3.

Rnd 9: Sl st in next dc; ch 3, dc in next 10 dc, in next ch-1 sp work (3 dc, ch 1, dc); ch 3, * sk next ch-3 lp and next dc, dc in next 11 dc, in next ch-1 sp work (3 dc, ch 1, dc); ch 3; rep from * 6 times more; join in 3rd ch of beg ch-3.

Rnd 10: Sl st in next dc; ch 3, dc in next 12 dc, in next ch-1 sp work (3 dc, ch 1, dc); ch 3, * sk next ch-3 lp and next dc, dc in next 13 dc, in next ch-1 sp work (3 dc, ch 1, dc); ch 3; rep from * 6 times more; join in 3rd ch of beg ch-3.

Rnd 11: Sl st in next dc; ch 3, dc in next 14 dc, 2 dc in next ch-1 sp; ch 5, * sk next ch-3 lp and next dc, dc in next 15 dc, 2 dc in next ch-1 sp; ch 5; rep from * 6 times more; join in 3rd ch of beg ch-3.

Rnd 12: Sl st in next dc; ch 3, dc in next 14 dc, * ch 2, sk next dc, 3 dc in 3rd ch of next ch-5 lp; ch 2, sk next dc, dc in next 15 dc; rep from * 7 times more, ending last rep without working last 15 dc; join in 3rd ch of beg ch-3.

Rnd 13: Sl st in next dc; ch 3, dc in next 12 dc, * ch 2, sk next dc, 3 dc in next dc; ch 1, sk next dc, 3 dc in next dc; ch 2, sk next dc, dc in next 13 dc; rep from * 7 times more, ending last rep without working last 13 dc; join in 3rd ch of beg ch-3.

Rnd 14: Sl st in next dc; ch 3, dc in next 10 dc, * ch 2, sk next dc, 3 dc in next dc; ch 1, 3 dc in next ch-1 sp; ch 1, sk next 2 dc, 3 dc in next dc; ch 2, sk next dc, dc in next 11 dc; rep from * 7 times more, ending last rep without working last 11 dc; join in 3rd ch of beg ch-3.

Rnd 15: Sl st in next dc; ch 3, dc in next 8 dc, * ch 2, sk next dc, 3 dc in next dc; ch 1, (3 dc in next ch-1 sp, ch 1) twice; sk next 2 dc, 3 dc in next dc; ch 2, sk next dc, dc in next 9 dc; rep from * 7 times more, ending last rep without working last 9 dc; join in 3rd ch of beg ch-3.

Rnd 16: Sl st in next dc; ch 3, dc in next 6 dc, * ch 2, sk next dc, 3 dc in next dc; ch 1, (3 dc in next ch-1 sp, ch 1) 3 times; sk next 2 dc, 3 dc in next dc; ch 2, sk next dc, dc in next 7 dc; rep from * 7 times more, ending last rep without working last 7 dc; join in 3rd ch of beg ch-3.

continued on page 39

Praline Pinwheel

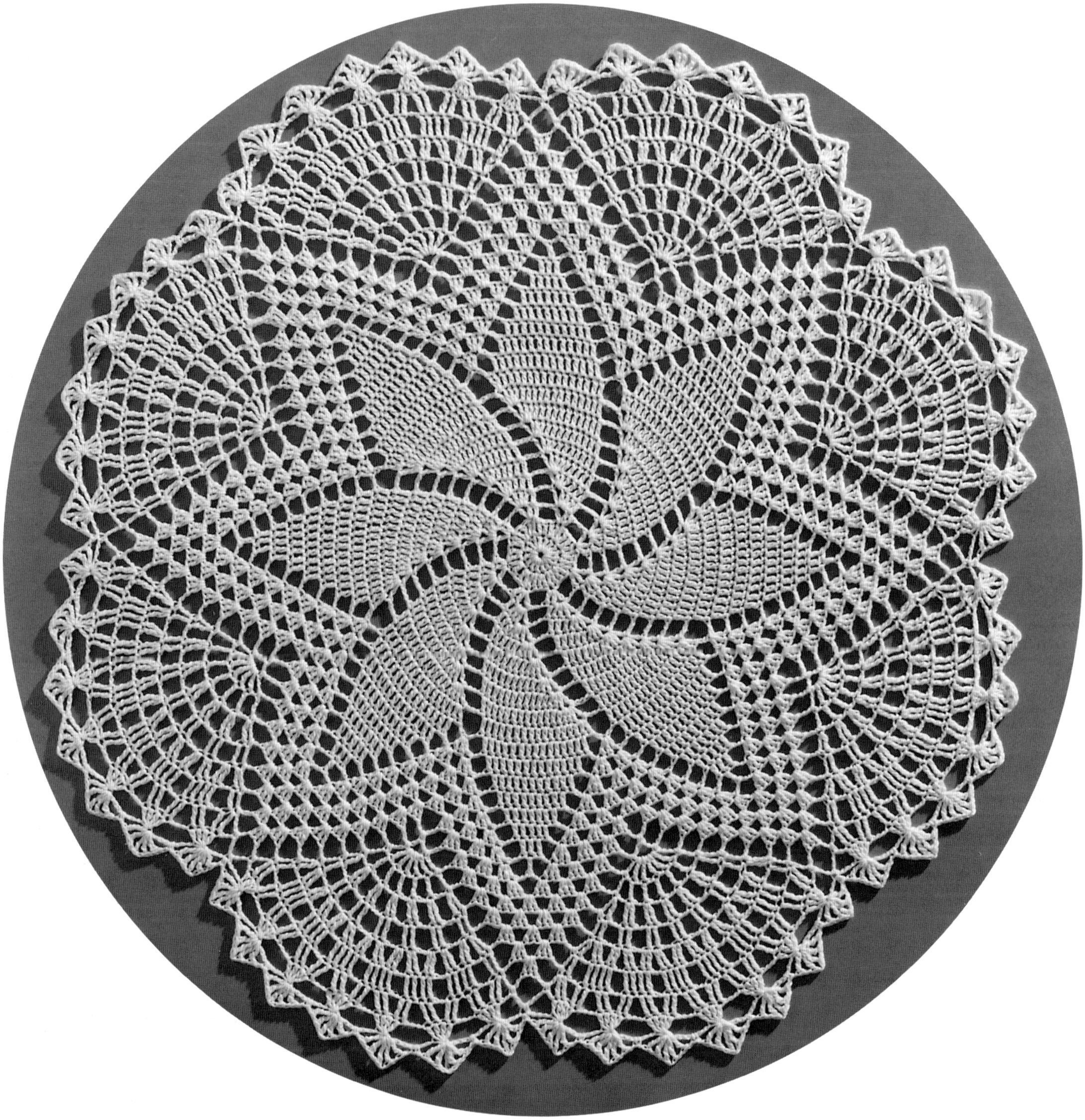

Rnd 17: Sl st in next dc; ch 3, dc in next 4 dc, * ch 2, sk next dc, 3 dc in next dc; (ch 1, 3 dc in next ch-1 sp) twice; ch 5, (3 dc in next ch-1 sp, ch 1) twice; sk next 2 dc, 3 dc in next dc; ch 2, sk next dc, dc in next 5 dc; rep from * 7 times more, ending last rep without working last 5 dc; join in 3rd ch of beg ch-3.

Rnd 18: Sl st in next dc; ch 3, dc in next 2 dc, * ch 2, sk next dc, 3 dc in next dc; (ch 1, 3 dc in next ch-1 sp) twice; ch 5, sc in next ch-5 lp, ch 5, (3 dc in next ch-1 sp, ch 1) twice; sk next 2 dc, 3 dc in next dc; ch 2, sk next dc, dc in next 3 dc; rep from * 7 times more, ending last rep without working last 3 dc; join in 3rd ch of beg ch-3.

Rnd 19: Sl st in next dc; ch 5 (counts as a dc and ch-2 sp), * sk next dc, 3 dc in next dc; (ch 1, 3 dc in next ch-1 sp) twice; ch 5, sc in next ch-5 lp, ch 3, sc in next ch-5 lp, ch 5, (3 dc in next ch-1 sp, ch 1) twice; sk next 2 dc, 3 dc in next dc; ch 2, sk next dc, dc in next dc, ch 2; rep from * 7 times more, ending last rep without working last dc and ch 2; join in 3rd ch of beg ch-5.

Rnd 20: Sl st in next 2 chs and in next dc; ch 3, 2 dc in same dc as last sl st made; * (ch 1, 3 dc in next ch-1 sp) twice; ch 5, sc in next ch-5 lp, ch 1, in next ch-3 lp work (trc, ch 1) 5 times; sc in next ch-5 lp, ch 5, (3 dc in next ch-1 sp, ch 1) twice; sk next 2 dc, 3 dc in next dc; ch 1, sk next dc, 3 dc in next dc; rep from * 7 times more, ending last rep without working last 3 dc; join in 3rd ch of beg ch-3.

Rnd 21: Sl st in next 2 dc and in next ch-1 sp; ch 3, 2 dc in same sp; ch 1, 3 dc in next ch-1 sp; * † ch 5, sc in next ch-5 lp, ch 2, (trc in next trc, ch 2) 5 times; sc in next ch-5 lp, ch 5 †, (3 dc in next ch-1 sp, ch 1) 4 times; 3 dc in next ch-1 sp; rep from * 6 times more, then rep from † to † once; (3 dc in next ch-1 sp, ch 1) 3 times; join in 3rd ch of beg ch-3.

Rnd 22: Sl st in next 2 dc and in next ch-1 sp; ch 3, 2 dc in same sp; * ch 5, sc in next ch-5 lp, trc in next sc, ch 2, (2 trc in next trc, ch 2) 5 times; trc in next sc, sc in next ch-5 lp, ch 5, (3 dc in next ch-1 sp, ch 1) 3 times; 3 dc in next ch-1 sp; rep from * 7 times more, ending last rep without working last 3 dc; join in 3rd ch of beg ch-3.

Rnd 23: Sl st in next 2 dc, in next 3 chs of next ch-5 lp, and in same lp; ch 1, sc in same lp, * trc in next trc, ch 2, (trc in next 2 trc, ch 2) 5 times; trc in next trc, sc in next ch-5 lp, ch 5, (3 dc in next ch-1 sp, ch 1) twice; 3 dc in next ch-1 sp; ch 5, sc in next ch-5 lp; rep from * 7 times more, ending last rep without working last sc; join in first sc.

Rnd 24: Ch 6 (counts as a trc and ch-2 sp on this and following rnd), * 2 trc in next trc; † 2 trc in next trc; ch 2, 2 trc in next trc †; rep from † to † 4 times more; 2 trc in next trc; ch 2, trc in next sc, sc in next ch-5 lp, ch 5, 3 dc in next ch-1 sp; ch 1, 3 dc in next ch-1 sp; ch 5, sc in next ch-5 lp, trc in next sc, ch 2; rep from * 7 times more, ending last rep without working last trc and ch 2; join in 4th ch of beg ch-6.

Rnd 25: Ch 6, * (trc in next 4 trc, ch 2) 6 times; trc in next trc, sc in next ch-5 lp, ch 5, 3 dc in next ch-1 sp; ch 5, sc in next ch-5 lp, trc in next trc, ch 2; rep from * 7 times more, ending last rep without working last trc and ch 2; join in 4th ch of beg ch-6.

Rnd 26: Ch 4 (counts as a trc on this and following rnds), * † trc in next 2 trc, ch 3, trc in next 2 trc †; rep from † to † 5 times more; trc in next trc, trc in next sc, ch 2, sc in next ch-5 lp, ch 6, sc in next ch-5 lp, ch 2, trc in next sc, trc in next trc; rep from * 7 times more, ending last rep without working last trc; join in 4th ch of beg ch-4.

Rnd 27: Sl st in next 2 trc, in next 3 chs, and in next trc; beg CL (see Pattern Stitches, page 37) over next 3 trc; [ch 7, CL (see Pattern Stitches. page 37) over next 4 trc] 5 times; ch 3, sc in next ch-6 lp, ch 3, CL over next 4 trc; * (ch 7, CL over next 4 trc) 6 times; ch 3, sc in next ch-6 lp, ch 3, CL over next 4 trc; rep from * 6 times more; ch 7; join in top of beg CL.

Rnd 28: Ch 5 (counts as a trc and a ch-1 sp), in top of same beg CL work (trc, ch 1) 4 times; sc in next ch-7 lp, ch 1, † in top of next CL work (trc, ch 1) 5 times; sc in next ch-7 lp, ch 1 †; rep from † to † 3 times more; in top of next CL work (trc, ch 1) 5 times; sc in next sc, ch 1; * rep from † to † 6 times; in top of next CL work (trc, ch 1) 5 times; sc in next sc, ch 1; rep from * 6 times more, then rep from † to † once; join in 4th ch of beg ch-5.

Finish off and weave in all ends.

Captain's Wheel

designed by Jo Ann Maxwell

Size:

About 16" diameter

Materials:

Bedspread-weight crochet cotton, one 225-yd ball ecru
Size 6 steel crochet hook, or size required for gauge

Gauge:

8 dc = 1"

Instructions

Ch 5, join to form a ring.

Rnd 1: Ch 3 (counts as a dc on this and following rnds), 19 dc in ring; join in 3rd ch of beg ch-3: 20 dc.

Rnd 2: Ch 6 (counts as a trc and a ch-2 sp), * trc in next dc, ch 2; rep from * 18 times more; join in 4th ch of beg ch-6.

Rnd 3: Ch 2 (counts as an hdc on this and following rnds), 2 hdc in same ch as joining; ch 5, * 3 hdc in next trc; ch 5; rep from * around; join in 2nd ch of beg ch-2.

Rnd 4: Sl st in next 2 hdc and in next 2 chs of next ch-5 lp; ch 2, 2 hdc in same ch as last sl st made; ch 3, sk next ch, 3 hdc in next ch; * 3 hdc in 2nd ch of next ch-5 lp; ch 3, sk next ch, 3 hdc in next ch; rep from * around; join in 2nd ch of beg ch-2.

Rnd 5: Sl st in next 2 hdc and in next ch-3 lp; ch 1, sc in same lp; ch 3, sl st in sc just made: picot made; ch 5, * sc in next ch-3 lp; ch 3, sl st in sc just made: picot made; ch 5; rep from * around; join in first sc.

Rnd 6: Sl st in next 2 chs of next ch-5 lp and in same lp; ch 1, sc in same lp; ch 7, * sc in next ch-5 lp, ch 7; rep from * around; join in first sc.

Rnd 7: Working around next ch-7 lp, sc in next ch, hdc in next ch, dc in next ch; 3 dc in next ch: tip made; dc in next ch, hdc in next ch, sc in next ch; * working around next ch-7 lp, sc in next ch, hdc in next ch, dc in next ch; 3 dc in next ch: tip made; dc in next ch, hdc in next ch, sc in next ch; rep from * around; join in first sc.

Rnd 8: Sl st in next 4 sts; ch 1, sc in same dc as last sl st made; ch 9, * sc in 2nd dc of tip of next ch-7 lp, ch 9; rep from * around; join in first sc.

Rnd 9: * Working around next ch-9 lp, sc in next ch, hdc in next ch, dc in next ch, 2 dc in each of next 2 chs; ch 22, sl st in 10th ch from hook; (ch 5, sk next 5 chs, sl st in next ch) twice: 3-lp group made; 2 dc in same ch as last 2 dc made; 2 dc in next ch; dc in next ch, hdc in next ch, sc in next ch; rep from * around; join in first sc: twenty 3-lp groups.

Rnd 10: Sl st in next 5 sts; * sl st in base of next 3-lp group; working around same 3-lp group, † sc in next ch, 2 hdc in each of next 3 chs; sc in next ch, sl st in next sl st †; rep from † to † once more; sc in next ch, 2 hdc in each of next 3 chs; 3 hdc in next ch; 2 hdc in each of next 3 chs; sc in next ch, sl st in next sl st; rep from † to † twice; ch 8, sl st in 4th ch from hook; ch 4; rep from * around; join in sl st at base of first 3-lp group. Finish off.

Rnd 11: Join thread in 5th st of top lp of any 3-lp group; ch 5 (counts as a dc and a ch-2 sp), sk next hdc, dc in next hdc, (ch 2, sk next hdc, dc in next hdc) 3 times; ch 5, * dc in 5th st of top lp of next 3-lp group, (ch 2, sk next hdc, dc in next hdc) 4 times; ch 5; rep from * around; join in 3rd ch of beg ch-5.

Rnd 12: Ch 3, dc in same ch as joining; (dc in next 2 chs, 2 dc in next dc) 4 times: scallop made; sc in next ch-5 lp, picot; * 2 dc in next dc; (dc in next 2 chs, 2 dc in next dc) 4 times: scallop made; sc in next ch 5 lp, picot; rep from * around; join in 3rd ch of beg ch-3.

Rnd 13: Working in bls only, sl st in next dc; ch 6 (counts as a trc and a ch-2 sp), * sk next dc, (trc in next dc, ch 2, sk next dc) 6 times; keeping last lp of each trc on hook, trc in next dc and in 2nd dc of next scallop, YO and draw through all 3 lps on hook: cluster made; ch 2; rep from * 18 times more; (trc in next dc, ch 2, sk next dc) 6 times; trc in next dc; join in 4th ch of beg ch-6.

Rnd 14: Sl st in next ch; ch 3, * working in bls only, dc in next 18 sts; keeping last lp of each dc on hook, dc in next ch, sk next cluster, dc in next ch; YO and draw through all 3 lps on hook; rep from * 18 times more; dc in next 19 sts; join in 3rd ch of beg ch-3.

Rnd 15: Ch 1, sc in same ch as joining; * sk next dc, working in bls only, 3 hdc in next dc; (ch 3, sk next 2 dc, 3 hdc in next dc) 5 times; sk next dc, sc in next st; rep from * around, ending last rep without working last sc; join in first sc.

Finish off and weave in all ends.

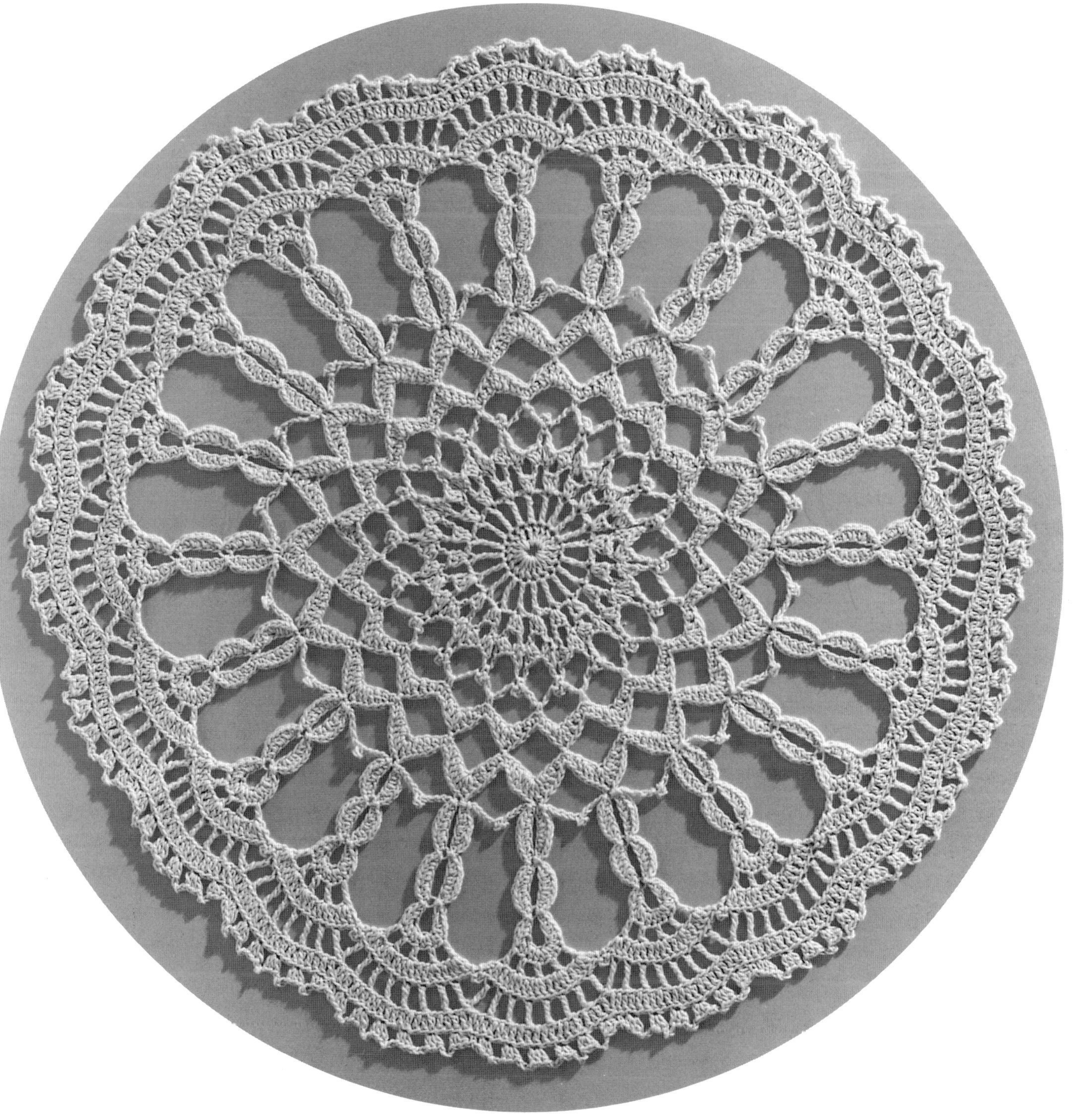

Infinity

designed by Jo Ann Maxwell

Size:
About 19" diameter

Materials:
Bedspread-weight crochet cotton, two 225-yd balls cream
Size 6 steel crochet hook, or size required for gauge

Gauge:
8 dc = 1"

Instructions
Ch 4.

Rnd 1: 11 dc in 4th ch from hook (3 skipped chs count as a dc); join in 3rd ch of beg ch-4: 12 dc.

Rnd 2: Ch 5 (counts as a trc and a ch-1 sp), trc in same ch as joining; ch 1, * in next dc work (trc, ch 1, trc); ch 1; rep from * 10 times more; join in 4th ch of beg ch-5: 24 trc.

Rnd 3: Ch 5 (counts as a dc and a ch-2 sp), sc in next trc; ch 3, sl st in sc just made: picot made; ch 2, * dc in next trc, ch 2, sc in next trc; ch 3, sl st in sc just made: picot made; ch 2; rep from * 10 times more; join in 3rd ch of beg ch-5.

Rnd 4: Ch 3 (counts as a dc on this and following rnds), in same ch as joining work (dc, ch 2, 2 dc): beg shell made; ch 3, * in next dc work (2 dc, ch 2, 2 dc): shell made; ch 3; rep from * 10 times more; join in 3rd ch of beg ch-3: 12 shells.

Rnd 5: Sl st in next dc and in next ch-2 sp; ch 6 (counts as a dc and a ch-3 lp), dc in same sp: beg V-st made; ch 3, sc in next ch-3 lp, picot; ch 3, * in ch-2 sp of next shell work (dc, ch 3, dc): V-st made; ch 3, sc in next ch-3 lp, picot; ch 3; rep from * 10 times more; join in 3rd ch of beg ch-6: 12 V-sts.

Rnd 6: Sl st in next ch-3 lp; ch 1, sc in same lp; picot; ch 11, * sc in ch-3 lp of next V-st, picot; ch 11; rep from * 10 times more; join in first sc.

Rnd 7: Sl st in next 3 chs of next ch-11 lp; ch 3, dc in same ch; * (ch 2, sk next ch, 2 dc in next ch) 3 times; ch 2, sk next picot, 2 dc in 3rd ch of next ch-11 lp; rep from * around, ending last rep without working last 2 dc; join in 3rd ch of beg ch-3.

Rnd 8: Sl st in next dc and in next ch-2 sp; ch 7 (counts as a trc and a ch-3 lp), * trc in next ch-2 sp, ch 3; rep from * around; join in 4th ch of beg ch-7.

Rnd 9: Beg V-st in same ch as joining; ch 3, sc in next trc, picot; ch 3, * V-st in next trc; ch 3, sc in next trc, picot; ch 3; rep from * around; join in 3rd ch of beg ch-6.

Rnd 10: Sl st in next ch-3 lp; ch 1, sc in same lp; picot; ch 9, * sc in ch-3 lp of next V-st, picot; ch 9; rep from * around; join in first sc.

Rnd 11: Sl st in next 5 chs of next ch-9 lp; beg shell in same ch; ch 6, * shell in 5th ch of next ch-9 lp; ch 6; rep from * around; join in 3rd ch of beg ch-3.

Rnd 12: Sl st in next dc and in next ch-2 sp; beg V-st in same sp; ch 3, sc in next ch-6 lp, picot; ch 3, * V-st in ch-2 sp of next shell; ch 3, sc in next ch-6 lp, picot; ch 3; rep from * around; join in 3rd ch of beg ch-6.

Rnd 13: Sl st in next ch-3 lp; sc in same lp; picot; ch 11, * sc in ch-3 lp of next V-st, picot; ch 11; rep from * around; join in first sc.

Rnds 14 through 16: Rep Rnds 7 through 9.

Rnd 17: Sl st in next ch-3 lp; ch 1, sc in same lp; picot; ch 9, * sc in ch-3 lp of next V-st, picot; ch 9; rep from * around; join in first sc.

Rnd 18: Sl st in next 5 chs of next ch-9 lp; ch 1, sc in same lp; picot; ch 9, * sc in next ch-9 lp, picot; rep from * around; join in first sc.

Rnds 19 through 23: Rep Rnd 18.

Rnds 24 through 26: Rep Rnds 11 through 13.

Rnd 27: Rep Rnd 7.

Finish off and weave in all ends.

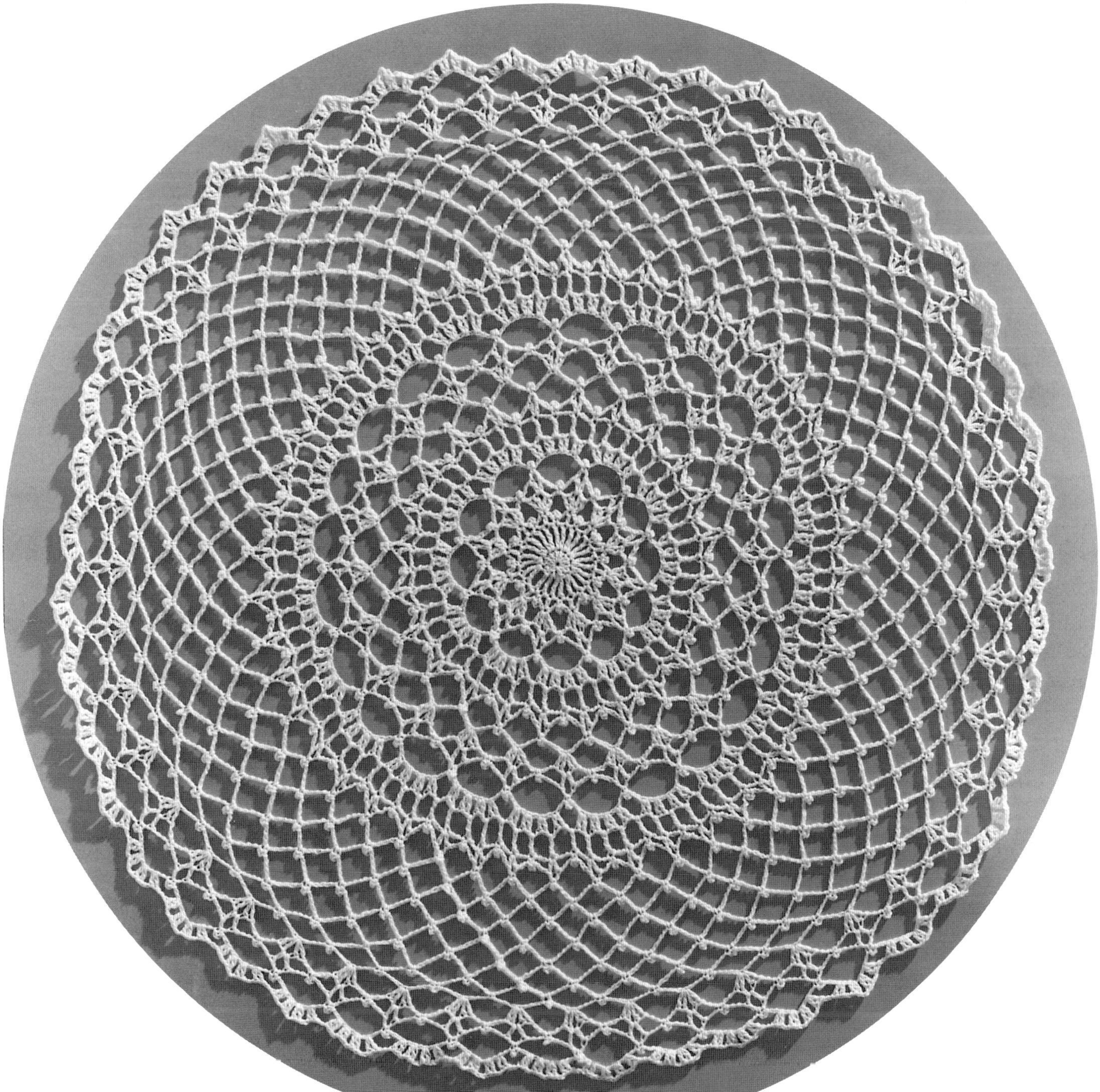

Crystal Fan

designed by Linda Mershon

Size:

About 17" at widest point

Materials:

Bedspread-weight crochet cotton, one 225-yd ball ecru
Size 6 steel crochet hook, or size required for gauge

Gauge:

8 dc = 1"

Pattern Stitches

Cluster (CL)**:** Keeping last lp of each dc on hook, dc in next 3 dc, YO and draw through all 4 lps on hook: CL made.
4-dc Cluster (4-dc CL)**:** Keeping last lp of each dc on hook, dc in next 4 dc, YO and draw through all 5 lps on hook: 4-dc CL made.
5-dc Cluster (5-dc CL)**:** Keeping last lp of each dc on hook, dc in next 5 dc, YO and draw through all 6 lps on hook: 5-dc CL made.
3-trc Cluster (3-trc CL)**:** Keeping last lp of each trc on hook, trc in next 3 trc, YO and draw through all 4 lps on hook: 3-trc CL made.

Instructions

Ch 3, join to form a ring.

Rnd 1: Ch 6 (counts as a dc and a ch-3 lp), * dc in ring, ch 3; rep from * 4 times more; join in 3rd ch of beg ch-6: 6 ch-3 lps.

Rnd 2: Sl st in next ch-3 lp; ch 3 (counts as a dc on this and following rnds), in same lp work (dc, ch 3, 2 dc): beg shell made; in each rem ch-3 lp work (2 dc, ch 3, 2 dc): shell made; join in 3rd ch of beg ch-3: 6 shells.

Rnd 3: Sl st in next dc and in next ch-3 lp; beg shell in same lp; ch 3, * in ch-3 lp of next shell work shell: shell in shell made; ch 3; rep from * 4 times more; join in 3rd ch of beg ch-3.

Rnd 4: Sl st in next dc and in next ch-3 lp; beg shell in same lp; * ch 3, dc in 2nd ch of next ch-3 lp, ch 3, shell in next shell; rep from * 5 times more, ending last rep without working last shell; join in 3rd ch of beg ch-3.

Rnd 5: Sl st in next dc and in next ch-3 lp; beg shell in same lp; * ch 3, sk next 2 dc, 3 dc in next dc; ch 3, shell in next shell; sk next 2 dc, 7 dc in next dc; shell in next shell; rep from * twice more, ending last rep without working last shell; join in 3rd ch of beg ch-3.

Rnd 6: Sl st in next dc and in next ch-3 lp; beg shell in same lp; * ch 3, sk next 2 dc, 2 dc in each of next 3 dc; ch 3, shell in next shell; sk next 2 dc, (dc in next dc, ch 1) 6 times; dc in next dc, shell in next shell; rep from * twice more, ending last rep without working last shell; join in 3rd ch of beg ch-3.

Rnd 7: Sl st in next dc and in next ch-3 lp; beg shell in same lp; * ch 3, sk next 2 dc, (sc in next dc, ch 3) 6 times; shell in next shell; sk next 2 dc, (dc in next dc, ch 2) 6 times; dc in next dc, shell in next shell; rep from * twice more, ending last rep without working last shell; join in 3rd ch of beg ch-3.

Rnd 8: Sl st in next dc and in next ch-3 lp; beg shell in same lp; * ch 3, sk next ch-3 lp, (sc in next ch-3 lp, ch 3) 5 times; shell in next shell; sk next 2 dc, (dc in next dc, ch 3) 6 times; dc in next dc, shell in next shell; rep from * twice more, ending last rep without working last shell; join in 3rd ch of beg ch-3.

Rnd 9: Sl st in next dc and in next ch-3 lp; beg shell in same lp; * ch 3, sk next ch-3 lp, (sc in next ch-3 lp, ch 3) 4 times; shell in next shell; sk next 2 dc, dc in next dc, ch 3, (2 dc in next dc, ch 3) 5 times; dc in next dc, sk next 2 dc, shell in next shell; rep from * twice more, ending last rep without working last shell; join in 3rd ch of beg ch-3.

Rnd 10: Sl st in next dc and in next ch-3 lp; beg shell in same lp; * ch 3, sk next ch-3 lp, (sc in next ch-3 lp, ch 3) 3 times; shell in next shell; ch 2, sk next 2 dc, 2 dc in next dc; ch 3, (2 dc in next dc, dc in next dc, ch 3) 5 times; 2 dc in next dc; ch 2, shell in next shell; rep from * twice more, ending last rep without working last shell; join in 3rd ch of beg ch-3.

Rnd 11: Sl st in next dc and in next ch-3 lp; beg shell in same lp; * ch 3, sk next ch-3 lp, (sc in next ch-3 lp, ch 3) twice; shell in next shell; ch 3, sk next 2 dc, 2 dc in next dc; dc in next dc, ch 3, (2 dc in next dc, dc in next 2 dc, ch 3) 5 times; 2 dc in next dc; dc in next dc, ch 3, shell in next shell; rep from * twice more, ending last rep without working last shell; join in 3rd ch of beg ch-3.

continued on page 46

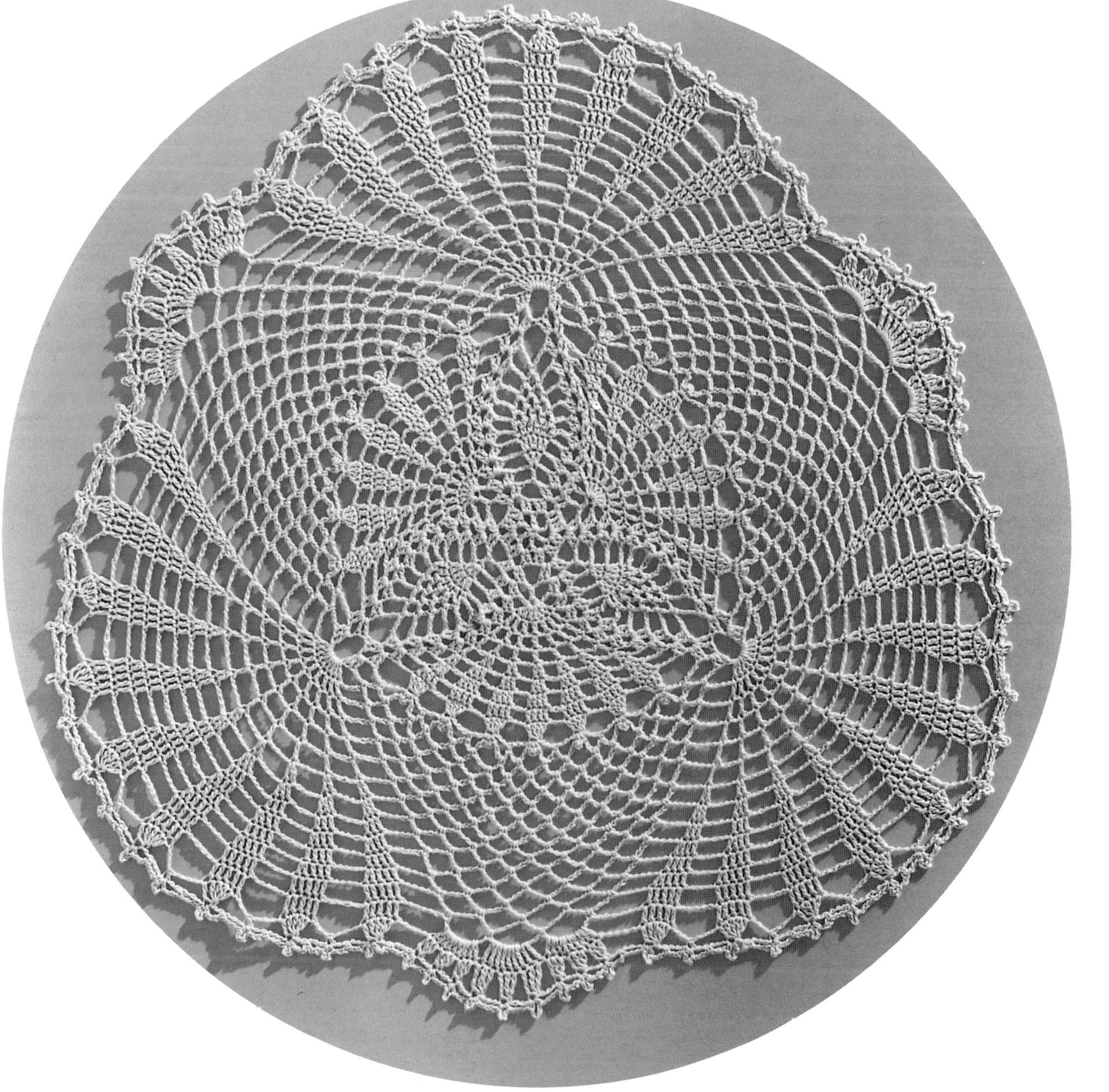

Rnd 12: Sl st in next dc and in next ch-3 lp; beg shell in same lp; * ch 3, sk next ch-3 lp, sc in next ch-3 lp, ch 3, shell in next shell; ch 3, sk next 2 dc, 2 dc in next dc; dc in next 2 dc, ch 3, (dc in next 4 dc, ch 3) 5 times; 2 dc in next dc; dc in next 2 dc, ch 3, shell in next shell; rep from * twice more, ending last rep without working last shell; join in 3rd ch of beg ch-3.

Rnd 13: Sl st in next dc and in next ch-3 lp; beg shell in same lp; * shell in next shell; ch 7, sk next 2 dc, [4-dc CL (see Pattern Stitches) over next 4 dc, ch 7] 7 times; shell in next shell; rep from * twice more, ending last rep without working last shell; join in 3rd ch of beg ch-3.

Rnd 14: Sl st in next dc and in next ch-3 lp; ch 3, 2 dc in same lp; * ch 3, 3 dc in ch-3 lp of next shell; † ch 5, sc in 4th ch of next ch-7 lp, ch 5, in top of next 4-dc CL work (sc, ch 3, sc) †; rep from † to † 6 times more; ch 5, sc in 4th ch of next ch-7 lp, ch 5, 3 dc in ch-3 lp of next shell; rep from * twice more, ending last rep without working last 3 dc; join in 3rd ch of beg ch-3.

Rnd 15: Sl st in next 2 dc and in next ch-3 lp; ch 3, 8 dc in same lp; * ch 5, (sc in next ch-5 lp, ch 5) 16 times; 9 dc in next ch-3 lp; rep from * twice more, ending last rep without working last 9 dc; join in 3rd ch of beg ch-3.

Rnd 16: Ch 4 (counts as a dc and a ch-1 sp), * † dc in next dc, ch 1 †; rep from † to † 6 times more; dc in next dc, ch 5, sk next ch-5 lp, (sc in next ch-5 lp, ch 5) 15 times; dc in next dc, ch 1; rep from * twice more, ending last rep without working last dc and ch 1; join in 3rd ch of beg ch-4.

Rnd 17: Ch 5 (count as a dc and a ch-2 sp), * † dc in next dc, ch 2 †; rep from † to † 6 times more; dc in next dc, ch 5, sk next ch-5 lp, (sc in next ch-5 lp, ch 5) 14 times; dc in next dc, ch 2; rep from * twice more, ending last rep without working last dc and ch 2; join in 3rd ch of beg ch-5.

Rnd 18: Ch 6 (counts as a dc and a ch-3 lp), * † dc in next dc, ch 3 †; rep from † to † 6 times more; dc in next dc, ch 5, sk next ch-5 lp, (sc in next ch-5 lp, ch 5) 13 times; dc in next dc, ch 3; rep from * twice more, ending last rep without working last dc and ch 3; join in 3rd ch of beg ch-6.

Rnd 19: Ch 7 (counts as a dc and a ch-4 lp), * † dc in next dc, ch 4 †; rep from † to † 6 times more; dc in next dc, ch 5, sk next ch-5 lp, (sc in next ch-5 lp, ch 5) 12 times; dc in next dc, ch 4; rep from * twice more, ending last rep without working last dc and ch 4; join in 3rd ch of beg ch-7.

Rnd 20: Ch 8 (counts as a dc and a ch-5 lp), * † dc in next dc, ch 5 †; rep from † to † 6 times more; dc in next dc, ch 5, sk next ch-5 lp, (sc in next ch-5 lp, ch 5) 11 times; dc in next dc, ch 5; rep from * twice more, ending last rep without working last dc and ch 5; join in 3rd ch of beg ch-8.

Rnd 21: Ch 3, dc in same ch as joining; * ch 5, (2 dc in next dc, ch 5) 8 times; sk next ch-5 lp, (sc in next ch-5 lp, ch 5) 10 times; 2 dc in next dc; rep from * twice more, ending last rep without working last 2 dc; join in 3rd ch of beg ch-3.

Rnd 22: Ch 3, dc in same ch as joining; dc in next dc, * ch 5, (2 dc in next dc, dc in next dc, ch 5) 8 times; sk next ch-5 lp, (sc in next ch-5 lp, ch 5) 9 times; 2 dc in next dc; dc in next dc; rep from * twice more, ending last rep without working last 3 dc; join in 3rd ch of beg ch-3.

Rnd 23: Ch 3, dc in next 2 dc, * ch 5, (dc in next 3 dc, ch 5) 8 times; sk next ch-5 lp, (sc in next ch-5 lp, ch 6) 8 times; sc in next ch-5 lp, ch 5, dc in next 3 dc; rep from * twice more, ending last rep without working last 3 dc; join in 3rd ch of beg ch-3.

Rnd 24: Ch 3, dc in same ch as joining and in next 2 dc; * ch 5, (2 dc in next dc, dc in next 2 dc, ch 5) 8 times; (sc in next ch-6 lp, ch 6) 6 times; sc in next ch-6 lp, ch 5, 2 dc in next dc; dc in next 2 dc; rep from * twice more, ending last rep without working last 4 dc; join in 3rd ch of beg ch-3.

Rnd 25: Ch 3, dc in next 3 dc, * ch 5, (dc in next 4 dc, ch 5) 8 times; (sc in next ch-6 lp, ch 6) 5 times; sc in next ch-6 lp, ch 5, dc in next 4 dc; rep from * twice more, ending last rep without working last 4 dc; join in 3rd ch of beg ch-3.

Rnd 26: Ch 3, dc in same ch as joining and in next 3 dc; * ch 5, (2 dc in next dc, dc in next 3 dc, ch 5) 8 times; (sc in next ch-6 lp, ch 7) 4 times; sc in next ch-6 lp, ch 5, 2 dc in next dc; dc in next 3 dc; rep from * twice more, ending last rep without working last 5 dc; join in 3rd ch of beg ch-3.

Rnd 27: Ch 3, dc in next 4 dc, * ch 6, (dc in next 5 dc, ch 6) 8 times; (sc in next ch-7 lp, ch 7) 3 times; sc in next ch-7 lp, ch 6, dc in next 5 dc; rep from * twice more, ending last rep without working last 5 dc; join in 3rd ch of beg ch-3.

Rnd 28: Ch 3, dc in next 4 dc, * ch 7, (dc in next 5 dc, ch 7) 7 times; dc in next 5 dc, ch 5, 9 dc in next ch-7 lp; 9 trc in next ch-7 lp; 9 dc in next ch-7 lp; ch 5, dc in next 5 dc; rep from * twice more, ending last rep without working last 5 dc; join in 3rd ch of beg ch-3.

Rnd 29: Ch 2, 4-dc CL; * † ch 6, sc in 4th ch of next ch-7 lp, ch 6, 5-dc CL (see Pattern Stitches, page 44) over next 5 dc †; rep from † to † 7 times more; ch 6, [CL (see Pattern Stitches, page 44) over next 3 dc, ch 3] twice; CL over next 3 dc; [ch 4, 3-trc CL (see Pattern Stitches, page 44) over next 3 trc] 3 times; ch 4, (CL over next 3 dc, ch 3) twice; CL over next 3 dc; ch 6, 5-dc CL over next 5 dc; rep from * twice more, ending last rep without working last 5-dc CL; join in top of 4-dc CL.

Rnd 30: Ch 1, in top of same 4-dc CL as joining work (sc, ch 3, sc); * † ch 3, sc in 3rd ch of next ch-6 lp, ch 3, in next sc work (sc, ch 3, sc); ch 3, sc in 3rd ch of next ch-6 lp, ch 3, in top of next CL work (sc, ch 3, sc) †; rep from † to † 7 times more; ch 3, sc in 3rd ch of next ch-6 lp, [ch 3, in top of next CL work (sc, ch 3, sc)] 9 times; ch 3, sc in 3rd ch of next ch-6 lp, ch 3, in top of next CL work (sc, ch 3, sc); rep from * twice more, ending last rep without working last (sc, ch 3, sc); join in first sc.

Finish off and weave in ends.

Eternal Love

designed by Jo Ann Maxwell

Size:
About 19" diameter

Materials:
Bedspread-weight crochet cotton, two 225-yd balls white
Size 6 steel crochet hook, or size required for gauge

Gauge:
8 dc = 1"

Pattern Stitches

Lover's Knot (LK): Draw up lp on hook 1/2", YO and draw through beg 1/2" lp just made, insert hook in lp on left side (not through beg 1/2" lp) as in **Fig 1**, YO and draw through, YO and draw through 2 lps on hook: LK made.

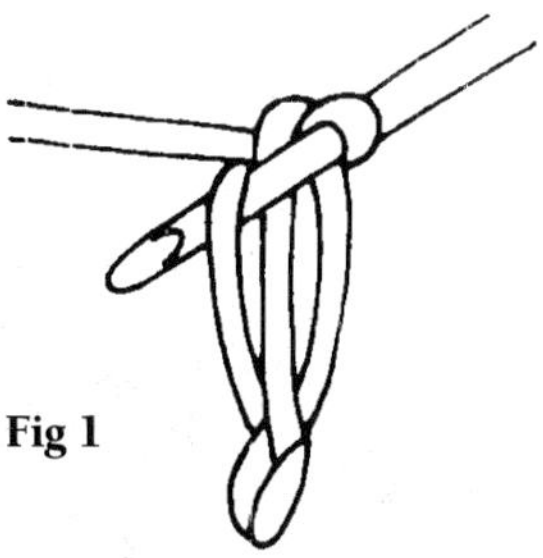

Fig 1

Double Lover's Knot (dLK): * Draw up lp on hook 3/8", YO and draw through beg 3/8" lp just made, insert hook in lp on left side (not through beg 3/8" lp) as in **Fig 1**, YO and draw through, YO and draw through 2 lps on hook; rep from * once more: dLK made.

Instructions

Ch 5; join to form a ring.

Rnd 1: Ch 3 (counts as a dc on this and following rnds), 19 dc in ring; join in 3rd ch of beg ch-3: 20 dc.

Rnd 2: Ch 6 (counts as a trc and a ch-2 sp), * trc in next dc, ch 2; rep from * around; join in 4th ch of beg ch-6.

Rnd 3: Ch 1, sc in same ch as joining; LK (see Pattern Stitches); * sc in next trc, LK; rep from * around; join in first sc.

Rnd 4: Ch 1, sc in same sc as joining; ch 5, * sc in next sc, ch 5; rep from * around; join in first sc.

Rnd 5: Sl st in next 2 chs of next ch-5 lp and in same lp; ch 1, sc in same lp; ch 3, sl st in sc just made: picot made; ch 5, * sc in next ch-5 lp; ch 3, sl st in sc just made: picot made; ch 5; rep from * around; join in first sc. Finish off.

Rnd 6: Join thread in any ch-5 lp; ch 1, sc in same lp as joining; dLK (see Pattern Stitches); * sc in next ch-5 lp, dLK; rep from * around; join in first sc. Finish off.

Rnd 7: Join thread in center knot of any dLK; ch 1, sc in same knot as joining; picot; ch 5, * sc in center knot of next dLK, picot; ch 5; rep from * around; join in first sc.

Rnd 8: Sl st in next 2 chs of next ch-5 lp; ch 3, dc in same ch as last sl st made; in next ch work (dc, ch 3, dc): V-st made; 2 dc in next ch; ch 2, * 2 dc in 2nd ch of next ch-5 lp; in next ch work (dc, ch 3, dc): V-st made; 2 dc in next ch; ch 2; rep from * around; join in 3rd ch of beg ch-3.

Rnd 9: Sl st in next 2 dc and in next ch-3 lp; ch 1, sc in same lp; picot; ch 9, * sc in ch-3 lp of next V-st, picot; ch 9; rep from * around; join in first sc.

Rnd 10: Sl st in next 2 chs of next ch-9 lp; ch 6 (counts as a dc and a ch-3 lp), * † sk next ch, dc in next ch, ch 3 †; rep from † to † once more; sk next ch; keeping last lp of each dc on hook, dc in next ch and in 2nd ch of next ch-9 lp; YO and draw through 3 lps on hook: cluster made; ch 3; rep from * 18 times more; then rep from † to † twice; sk next ch, dc in next dc; join in 3rd ch of beg ch-6.

Rnd 11: Ch 2 (counts as an hdc), hdc in each ch, in each dc, and in top of each cluster; join in 2nd ch of beg ch-2: 240 hdc.

Rnd 12: Ch 1, sc in same ch as joining; * LK; sk next 3 hdc, sc in bl of next hdc; rep from * around, ending last rep without working last sc; join in first sc.

Rnd 13: Ch 1, sc in same sc as joining; ch 5, * sc in next sc, ch 5; rep from * around; join in first sc.

Rnd 14: Sl st in next 2 chs of next ch-5 lp and in same lp; ch 1, sc in same lp; LK; * sc in next ch-5 lp, LK; rep from * around; join in first sc.

Rnds 15 and 16: Rep Rnds 13 and 14.

Rnd 17: Rep Rnd 13.

Rnd 18: Sl st in next 2 chs of next ch-5 lp; ch 3, dc in same ch; V-st in next ch; 2 dc in next ch; ch 2, sc in next ch-5 lp, picot; ch 2, * 2 dc in 2nd ch of next ch-5 lp; V-st in next ch; 2 dc in next ch; ch 2, sc in next ch-5 lp, picot; ch 2; rep from * around; join in 3rd ch of beg ch-3.

continued on page 50

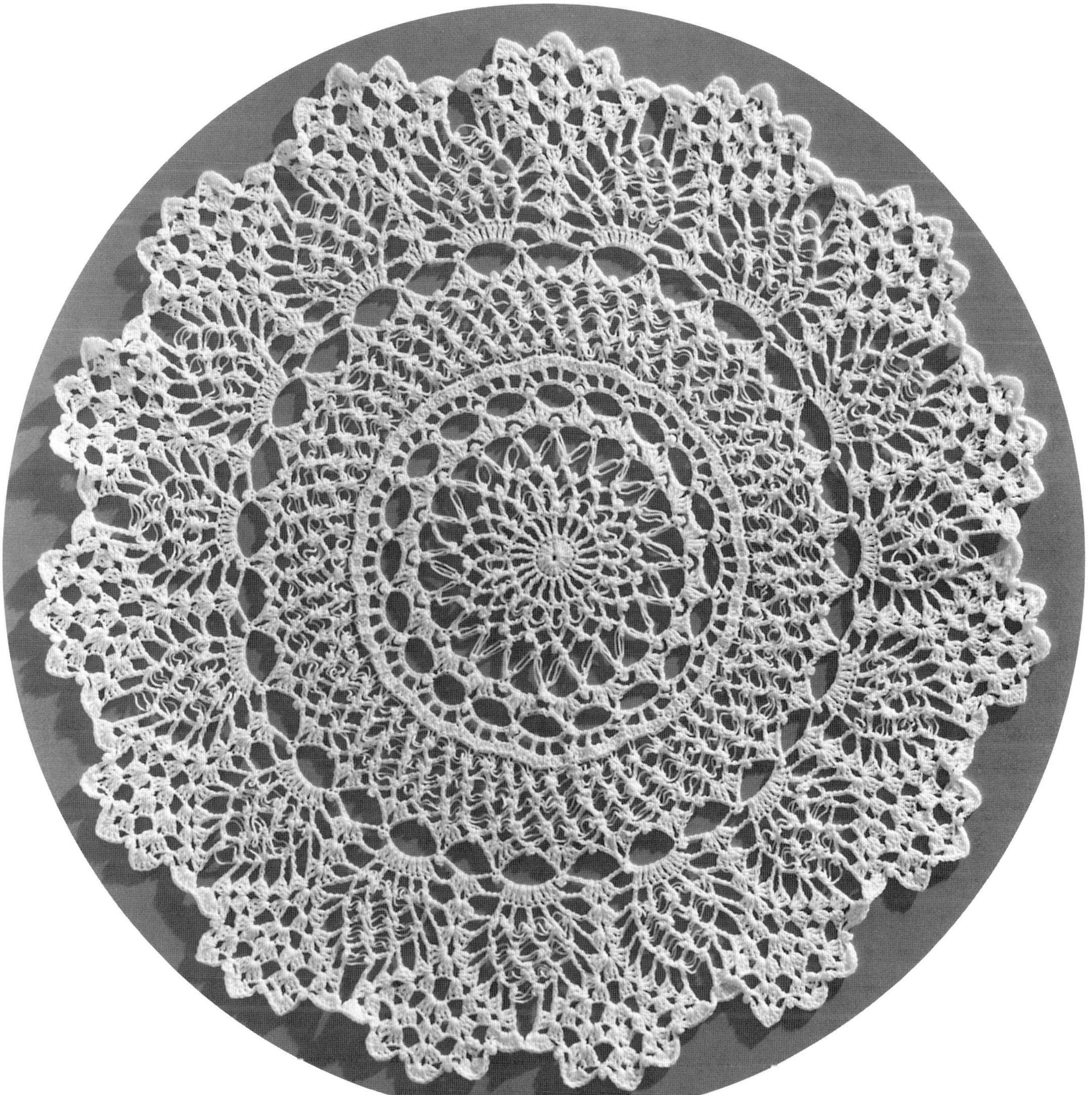

Rnd 19: Sl st in next 2 dc and in next ch-3 lp; ch 1, sc in same lp; picot; ch 9, * sc in ch-3 lp of next V-st, picot; ch 9; rep from * around; join in first sc.

Rnd 20: Sl st in next 2 chs of next ch-9 lp; ch 3, dc in same ch; (dc in next ch, 2 dc in next ch) 3 times: pineapple base made; ch 3, 2 dc in 4th ch of next ch-9 lp; V-st in next ch; 2 dc in next ch; ch 3, * 2 dc in 2nd ch of next ch-9 lp; (dc in next ch, 2 dc in next ch) 3 times: pineapple base made; ch 3, 2 dc in 4th ch of next ch-9 lp; V-st in next ch; 2 dc in next ch; ch 3; rep from * around; join in 3rd ch of beg ch-3: 15 pineapple bases.

Rnd 21: Sl st in next dc; ch 3, LK; (sk next dc, dc in next dc, LK) 3 times; sk next dc, dc in next dc, ch 3, in ch-3 lp of next V-st work (3 dc, ch 3, 3 dc): shell made; ch 3, * dc in 2nd dc of next pineapple base, LK; (sk next dc, dc in next dc, LK) 3 times; sk next dc, dc in next dc, ch 3, in ch-3 lp of next V-st work (3 dc, ch 3, 3 dc): shell made; ch 3; rep from * around; join in 3rd ch of beg ch-3.

Rnd 22: Ch 1, sc in same ch as joining; (ch 5, sc in next dc) 4 times; ch 3, in ch-3 lp of next shell work shell: shell in shell made; ch 3, sk next ch-3 lp, * sc in next dc, (ch 5, sc in next dc) 4 times; ch 3, in ch-3 lp of next shell work shell: shell in shell made; ch 3, sk next ch-3 lp; rep from * around; join in first sc.

Rnd 23: Sl st in next 2 chs of next ch-5 lp and in same lp; ch 1, sc in same lp; (LK, sc in next ch-5 lp) 3 times; ch 3, in ch-3 lp of next shell work (3 dc, ch 3) 3 times; * sc in next ch-5 lp, (LK, sc in next ch-5 lp) 3 times; ch 3, in ch-3 lp of next shell work (3 dc, ch 3) 3 times; rep from * around; join in first sc.

Rnd 24: Ch 1, sc in same sc as joining; (ch 5, sc in next sc) 3 times; ch 3, sk next ch-3 lp, (shell in next ch-3 lp, ch 3) twice; * sc in next sc, (ch 5, sc in next sc) 3 times; ch 3, sk next ch-3 lp, (shell in next ch-3 lp, ch 3) twice; rep from * around; join in first sc.

Rnd 25: Sl st in next 2 chs of next ch-5 lp and in same lp; ch 1, sc in same lp; * (LK, sc in next ch-5 lp) twice; ch 3, shell in next shell; ch 3, shell in next ch-3 lp; ch 3, shell in next shell; ch 3, sc in next ch-5 lp; rep from * around, ending last rep without working last sc; join in first sc.

Rnd 26: Ch 1, sc in same sc as joining; * (ch 5, sc in next sc) twice; ch 3, (shell in next shell, ch 1, shell in next ch-3 lp, ch 1) twice; shell in next shell; ch 3, sc in next sc; rep from * around, ending last rep without working last sc; join in first sc.

Rnd 27: Sl st in next 2 chs of next ch-5 lp and in same lp; ch 1, sc in same lp; * LK; sc in next ch-5 lp, ch 3, (shell in next shell, ch 2) 4 times; shell in next shell; ch 3, sc in next ch-5 lp; rep from * around, ending last rep without working last sc; join in first sc. Finish off.

Rnd 28: Join thread in ch-3 lp of first shell to left of any pineapple; ch 3, 6 dc in same lp; † ch 2, sc in next ch-2 sp, picot; ch 2, 7 dc in ch-3 lp of next shell †; rep from † to † 3 times more; * 7 dc in ch-3 lp of next shell; rep from † to † 4 times; rep from * around; join in 3rd ch of beg ch-3.

Finish off and weave in all ends.

Flowering Mum

designed by Yalanda Wiese

Size:

About 17" diameter

Materials:

Bedspread-weight crochet cotton, two 225-yd balls white
Size 6 steel crochet hook, or size required for gauge

Gauge:

8 dc = 1"

Instructions

Ch 5, join to form a ring.

Rnd 1: Ch 8 (counts as a trc and a ch-4 lp), * trc in ring, ch 4; rep from * 8 times more; join in 4th ch of beg ch-8: 10 ch-4 lps.

Rnd 2: Ch 3 (counts as a dc on this and following rnds), in same ch as joining work (dc, ch 2, 2 dc): beg shell made; sc in next ch-4 lp, * in next trc work (2 dc, ch 2, 2 dc): shell made; sc in next ch-4 lp; rep from * 8 times more; join in 3rd ch of beg ch-3: 10 shells.

Rnd 3: Sl st in next dc and in next ch-2 sp; beg shell in same sp; ch 1, * in ch-2 sp of next shell work shell: shell in shell made; ch 1; rep from * 8 times more; join in 3rd ch of beg ch-3.

Rnd 4: Sl st in next dc and in next ch-2 sp; beg shell in same sp; ch 1, dc in next ch-1 sp, ch 1, * shell in next shell; ch 1, dc in next ch-1 sp, ch 1; rep from * 8 times more; join in 3rd ch of beg ch-3.

Rnd 5: Sl st in next dc and in next ch-2 sp; beg shell in same sp; ch 2, dc in next dc, ch 2, * shell in next shell; ch 2, dc in next dc, ch 2; rep from * 8 times more; join in 3rd ch of beg ch-3.

Rnd 6: Sl st in next dc and in next ch-2 sp; beg shell in same sp; ch 3, dc in next dc, ch 3, * shell in next shell; ch 3, dc in next dc, ch 3; rep from * 8 times more; join in 3rd ch of beg ch-3.

Rnd 7: Sl st in next dc and in next ch-2 sp; beg shell in same sp; ch 4, dc in next dc, ch 4, * shell in next shell; ch 4, dc in next dc, ch 4; rep from * 8 times more; join in 3rd ch of beg ch-3.

Rnd 8: Sl st in next dc and in next ch-2 sp; beg shell in same sp; ch 5, dc in next dc, ch 5, * shell in next shell; ch 5, dc in next dc, ch 5; rep from * 8 times more; join in 3rd ch of beg ch-3.

Rnd 9: Sl st in next dc and in next ch-2 sp; ch 3, in same sp work (dc, ch 2, 2 dc, ch 2, 2 dc); * † ch 1, sc in next ch-5 lp close to dc of prev rnd, sc in next dc and in next ch-5 lp; ch 4 †, in next ch-2 sp work (2 dc, ch 2, 2 dc, ch 2, 2 dc); rep from * 7 times more, then rep from † to † once; join in 3rd ch of beg ch-3.

Rnd 10: Sl st in next dc and in next ch-2 sp; beg shell in same sp; * ch 1, shell in next ch-2 sp; ch 3, sc in next 3 sc, ch 3, shell in next ch-2 sp; rep from * 9 times more, ending last rep without working last shell; join in 3rd ch of beg ch-3.

Rnd 11: Sl st in next dc and in next ch-2 sp; beg shell in same sp; * ch 1, dc in next ch-1 sp, ch 1, shell in next shell; ch 3, sc in next 3 sc, ch 3, shell in next shell; rep from * 9 times more, ending last rep without working last shell; join in 3rd ch of beg ch-3.

Rnd 12: Sl st in next dc and in next ch-2 sp; beg shell in same sp; * ch 3, dc in next dc, ch 3, shell in next shell; ch 3, sk next sc, sc in next sc, ch 3, shell in next shell; rep from * 9 times more, ending last rep without working last shell; join in 3rd ch of beg ch-3.

Rnd 13: Sl st in next dc and in next ch-2 sp; beg shell in same sp; * ch 5, dc in next dc, ch 5, shell in next shell; ch 1, shell in next shell; rep from * 9 times more, ending last rep without working last shell; join in 3rd ch of beg ch-3.

Rnd 14: Sl st in next dc and in next ch-2 sp; beg shell in same sp; * ch 4, sc in next ch-5 lp close to dc of prev rnd, sc in next dc, sc in next ch-5 lp, ch 4, shell in next shell; shell in next ch-1 sp; shell in next shell; rep from * 9 times more, ending last rep without working last shell; join in 3rd ch of beg ch-3.

Rnd 15: Sl st in next dc and in next ch-2 sp; beg shell in same sp; * ch 4, sc in next 3 sc, ch 4, (shell in next shell, ch 1) twice; shell in next shell; rep from * 9 times more, ending last rep without working last shell; join in 3rd ch of beg ch-3.

continued on page 53

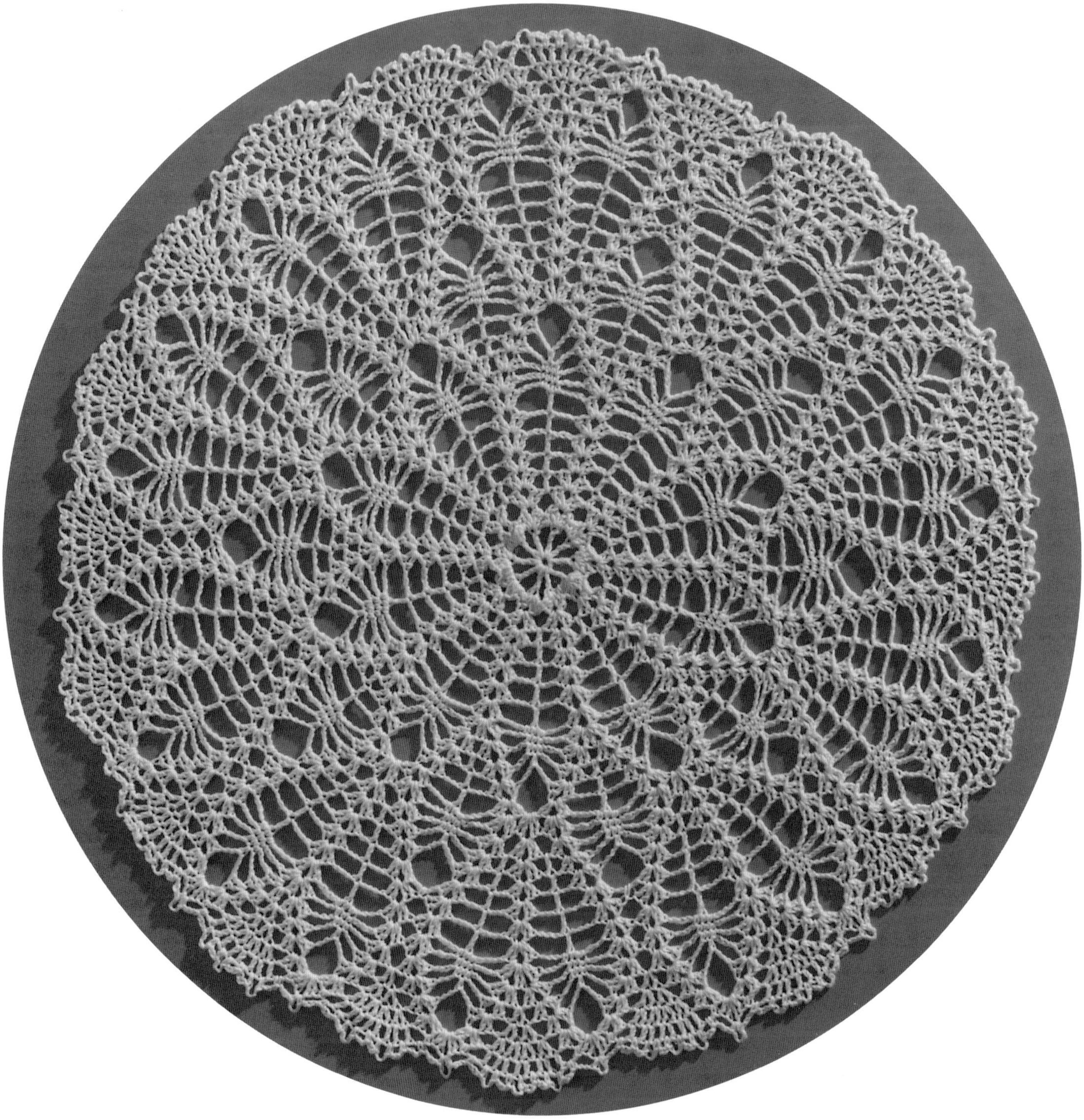

Rnd 16: Sl st in next dc and in next ch-2 sp; beg shell in same sp; * ch 3, sc in next 3 sc, ch 3, (shell in next shell, ch 1, dc in next ch-1 sp, ch 1) twice; shell in next shell; rep from * 9 times more, ending last rep without working last shell; join in 3rd ch of beg ch-3.

Rnd 17: Sl st in next dc and in next ch-2 sp; beg shell in same sp; * ch 3, sk next sc, sc in next sc, ch 3, (shell in next shell, ch 2, dc in next dc, ch 2) twice; shell in next shell; rep from * 9 times more, ending last rep without working last shell; join in 3rd ch of beg ch-3.

Rnd 18: Sl st in next dc and in next ch-2 sp; beg shell in same sp; * ch 4, (shell in next shell, ch 3, dc in next dc, ch 3) twice; shell in next shell; rep from * 9 times more, ending last rep without working last shell; join in 3rd ch of beg ch-3.

Rnd 19: Sl st in next dc and in next ch-2 sp; beg shell in same sp; * sc in next ch-4 lp, (shell in next shell, ch 4, dc in next dc, ch 4) twice; shell in next shell; rep from * 9 times more, ending last rep without working last shell; join in 3rd ch of beg ch-3.

Rnd 20: Sl st in next dc and in next ch-2 sp; beg shell in same sp; * (shell in next shell, ch 5, dc in next dc, ch 5) twice; shell in next shell; rep from * 9 times more, ending last rep without working last shell; join in 3rd ch of beg ch-3.

Rnd 21: Sl st in next dc and in next ch-2 sp; beg shell in same sp; * shell in next shell; † ch 4, sc in next ch-5 lp close to dc of prev rnd, sc in next dc, sc in next ch-5 lp, ch 4 †, in next shell work (2 dc, ch 2, 2 dc, ch 2, 2 dc); rep from † to † once; shell in next shell; rep from * 9 times more, ending last rep without working last shell; join in 3rd ch of beg ch-3.

Rnd 22: Sl st in next dc and in next ch-2 sp; beg shell in same sp; ch 1, * shell in next shell; ch 4, sc in next 3 sc, ch 4, shell in next ch-2 sp; ch 1, shell in next ch-2 sp; ch 4, sc in next 3 sc, ch 4, shell in next shell; ch 1; rep from * 9 times more; join in 3rd ch of beg ch-3.

Rnd 23: Sl st in next dc and in next ch-2 sp; beg shell in same sp; ch 1, in next ch-1 sp work (dc, ch 1, dc): V-st made; ch 1, * † shell in next shell; ch 3, sc in next 3 sc, ch 3 †, shell in next shell; ch 1, in next ch-1 sp work (dc, ch 1, dc): V-st made; ch 1; rep from * 18 times more, then rep from † to † once; join in 3rd ch of beg ch-3.

Rnd 24: Sl st in next dc and in next ch-2 sp; beg shell in same sp; ch 1, V-st in each of next 2 dc; ch 1, * † shell in next shell; ch 3, sk next sc, sc in next sc, ch 3 †, shell in next shell; ch 1, V-st in each of next 2 dc; ch 1; rep from * 18 times more, then rep from † to † once; join in 3rd ch of beg ch-3.

Rnd 25: Sl st in next dc and in next ch-2 sp; beg shell in same sp; ch 1, V-st in each of next 4 dc; ch 1, * (shell in next shell, ch 1) twice; V-st in each of next 4 dc; ch 1; rep from * 18 times more; shell in next shell; ch 1; join in 3rd ch of beg ch-3.

Rnd 26: Sl st in next dc and in next ch-2 sp; ch 1, sc in same sp; * ch 2, sk next 2 dc, (dc in next dc, ch 1) 7 times; dc in next dc, ch 2, sc in ch-2 sp of next shell, shell in next ch-1 sp; sc in ch-2 sp of next shell; rep from * 18 times more, ending last rep without working last sc; join in first sc.

Rnd 27: Sl st in next ch-2 sp; ch 1, sc in same sp; * (ch 3, sc in next ch-1 sp) 7 times; ch 3, sc in next ch-2 sp, ch 3, in ch-2 sp of next shell work (sc, ch 3, sc); ch 3, sc in next ch-2 sp; rep from * 18 times more, ending last rep without working last sc; join in first sc.

Finish off and weave in all ends.

Curtain Call

designed by Lucille LaFlamme

Size:
About 16" diameter

Materials:
Bedspread-weight crochet cotton, one (225 yd ball) ecru
Size 7 steel crochet hook, or size required for gauge

Gauge:
Rnds 1 through 5 = 3 1/4" diameter

Pattern Stitches

Cluster (CL)**:** Keeping last lp of each dc on hook, 2 dc in next lp or in next 2 sts; YO and draw through all 3 lps on hook: CL made.

Beg 4-dc Cluster (beg 4-dc CL)**:** Ch 3, keeping last lp of each dc on hook, dc in next 3 sts, YO and draw through all 4 lps on hook: beg 4-dc CL made.

4-dc CLuster (4-dc CL): Keeping last lp of each dc on hook, dc in next 4 sts, YO and draw through all 5 lps on hook:
4-dc CL made.

Instructions

Ch 9, join to form a ring.

Rnd 1: Ch 3 (counts as a dc on this and following rnds), 23 dc in ring; join in 3rd ch of beg ch-3: 24 dc.

Rnd 2: Ch 4 (counts as a dc and a ch-1 sp), * dc in next dc, ch 1; rep from * 22 times more; join in 3rd ch of beg ch-4.

Rnd 3: Ch 5 (counts as a dc and a ch-2 sp), * dc in next dc, ch 2; rep from * around; join in 3rd ch of beg ch-5.

Rnd 4: Ch 6 (counts as a dc and a ch-3 lp on this and following rnds), * dc in next dc, ch 3; rep from * around; join in 3rd ch of beg ch-6.

Rnd 5: Ch 6, * sc in next dc; ch 4, sl st in sc just made: picot made; ch 3, dc in next dc, ch 3; rep from * 11 times more, ending last rep without working last dc and last ch 3; join in 3rd ch of beg ch-6.

Rnd 6: Ch 12 (counts as a dc and a ch-9 lp), * dc in next dc, ch 9; rep from * 10 times more; join in 3rd ch of beg ch-12.

Rnd 7: Ch 8 (counts as a dc and a ch-5 lp), * sk next 4 chs of next ch-9 lp, dc in next ch, ch 5, dc in next dc, ch 5; rep from * 11 times more, ending last rep without working last dc and last ch 5; join in 3rd ch of beg ch-8.

Rnd 8: Ch 6, * sc in next ch-5 lp, picot; ch 3, dc in next dc, ch 3; rep from * 23 times more, ending last rep without working last dc and last ch 3; join in 3rd ch of beg ch-6.

Rnd 9: Ch 10 (counts as a dc and a ch-7 lp), * dc in next dc, ch 7; rep from * 22 times more; join in 3rd ch of beg ch-10.

Rnd 10: Ch 11 (counts as a dc and a ch-8 lp), * dc in next dc, ch 8; rep from * 22 times more; join in 3rd ch of beg ch-11.

Rnd 11: Sl st in next ch-8 lp; ch 3, 9 dc in same lp; * dc in next dc, 10 dc in next ch-8 lp; ch 2, 10 dc in next ch-8 lp; rep from * 11 times more, ending last rep without working last 10 dc; join in 3rd ch of beg ch-3.

Rnd 12: Ch 3, dc in next 4 dc, * ch 4, sk next 5 dc, in next dc work (dc, ch 2, dc): V-st made; ch 4, sk next 5 dc, dc in next 5 dc, ch 2, dc in next 5 dc; rep from * 11 times more, ending last rep without working last 5 dc; join in 3rd ch of beg ch-3.

Rnd 13: Ch 3, dc in next 4 dc, * ch 4, sk next ch-4 lp, in next ch-2 sp of next V-st work V-st: V-st in V-st made; ch 4, sk next ch-4 lp, dc in next 5 dc, ch 3, dc in next 5 dc; rep from * 11 times more, ending last rep without working last 5 dc; join in 3rd ch of beg ch-3.

Rnd 14: Ch 3, dc in next 4 dc, * ch 4, sk next ch-4 lp, V-st in next V-st; ch 4, sk next ch-4 lp, dc in next 5 dc, ch 5, dc in next 5 dc; rep from * 11 times more, ending last rep without working last 5 dc; join in 3rd ch of beg ch-3.

Rnd 15: Ch 3, dc in next 4 dc, * ch 3, sk next ch-4 lp, V-st in next V-st; ch 3, sk next ch-4 lp, dc in next 5 dc, ch 7, dc in next 5 dc; rep from * 11 times more, ending last rep without working last 5 dc; join in 3rd ch of beg ch-3.

Rnd 16: Ch 3, dc in next 4 dc, * ch 2, sk next ch-3 lp, V-st in next V-st; ch 2, sk next ch-3 lp, dc in next 5 dc, ch 5, CL (see Pattern Stitches) in next ch-7 lp; ch 5, dc in next 5 dc; rep from * 11 times more, ending last rep without working last 5 dc; join in 3rd ch of beg ch-3.

Rnd 17: Ch 3, dc in next 4 dc, * ch 1, sk next ch-2 sp, V-st in next V-st; ch 1, sk next ch-2 sp, dc in next 5 dc, ch 5, CL in next ch-5 lp; ch 3, CL in next ch-5 lp; ch 5, dc in next 5 dc; rep from * 11 times more, ending last rep without working last 5 dc; join in 3rd ch of beg ch-3.

continued on page 56

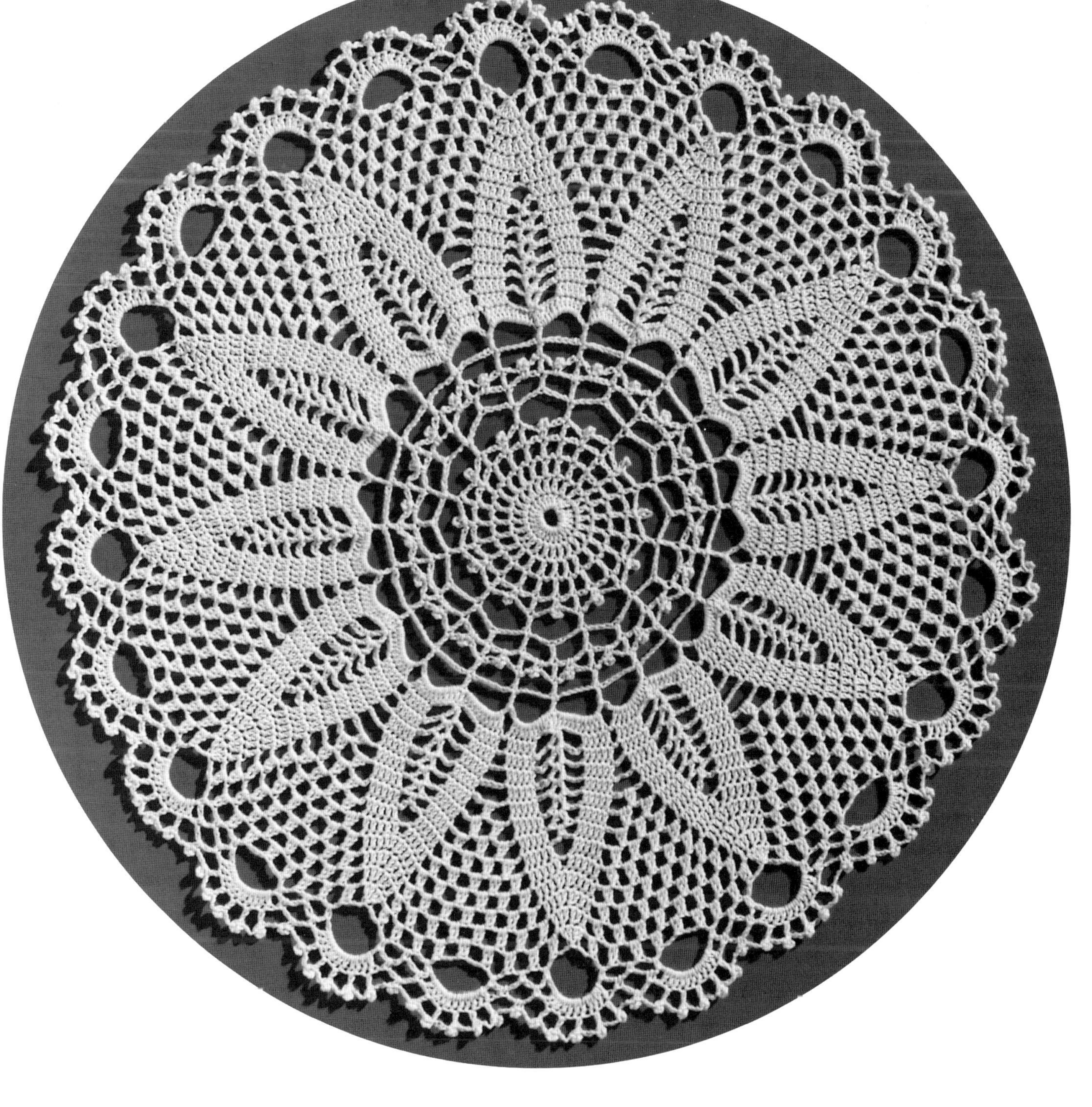

Rnd 18: Ch 3, dc in next 4 dc, * ch 3, sc in ch-2 sp of next V-st, ch 3, dc in next 5 dc, ch 5, CL in next ch-5 lp; ch 3, CL in next ch-3 lp; ch 3, CL in next ch-5 lp; ch 5, dc in next 5 dc; rep from * 11 times more, ending last rep without working last 5 dc; join in 3rd ch of beg ch-3.

Rnd 19: Ch 3, dc in next 4 dc, * ch 2, dc in next 5 dc, ch 3, CL in next ch-5 lp; (ch 3, CL in next ch-3 lp) twice; ch 3, CL in next ch-5 lp; ch 3, dc in next 5 dc; rep from * 11 times more, ending last rep without working last 5 dc; join in 3rd ch of beg ch-3.

Rnd 20: Ch 3, dc in next 4 dc, *sk next ch-2 sp, dc in next 5 dc, (ch 3, CL in next ch-3 lp) 5 times; ch 3, dc in next 5 dc; rep from * 11 times more, ending last rep without working last 5 dc; join in 3rd ch of beg ch-3.

Rnd 21: Ch 3, dc in next 2 dc, * (CL over next 2 dc) twice; dc in next 3 dc, (ch 3, CL in next ch-3 lp) 6 times; ch 3, dc in next 3 dc; rep from * 11 times more, ending last rep without working last 3 dc; join in 3rd ch of beg ch-3.

Rnd 22: Ch 3, dc in next dc, * (CL over next 2 sts) twice; dc in next 2 dc, (ch 3, CL in next ch-3 lp) 7 times; ch 3, dc in next 2 dc; rep from * 11 times more, ending last rep without working last 2 dc; join in 3rd ch of beg ch-3.

Rnd 23: Ch 3, * (CL over next 2 sts) twice; dc in next dc, (ch 3, CL in next ch-3 lp) 8 times; ch 3, dc in next dc; rep from * 11 times more, ending last rep without working last dc; join in 3rd ch of beg ch-3.

Rnd 24: Beg 4-dc CL (see Pattern Stitches, page 53) over next 3 sts; * (ch 3, CL in next ch-3 lp) 9 times; ch 3, 4-dc CL (see Pattern Stitches, page 53) over next 4 sts; rep from * 11 times more, ending last rep without working last 4-dc CL; join in top of beg 4-dc CL.

Rnd 25: Sl st in next 3 chs, in top of next CL, and in next 2 chs of next ch-3 lp; ch 1, sc in same lp; * (ch 5, sc in next ch-3 lp) twice; ch 11, sk next 2 ch-3 lps, sc in next ch-3 lp; rep from * 23 times more, ending last rep without working last sc; join in first sc.

Rnd 26: Sl st in next 3 chs of next ch-5 lp; ch 1, sc in same lp; * ch 5, sc in next ch-5 lp, ch 1, 11 dc in next ch-11 lp; ch 1, sc in next ch-5 lp; rep from * 23 times more, ending last rep without working last sc; join in first sc.

Rnd 27: Sl st in next 3 chs of next ch-5 lp; ch 1, sc in same lp; picot; * (ch 3, sk next dc, dc in next dc, picot) 5 times; ch 3, sc in next ch-5 lp, picot; rep from * 23 times more, ending last rep without working last sc and last picot; join in first sc.

Finish off and weave in ends.

On Broadway

designed by Lucille LaFlamme

Size:

About 15" diameter

Materials:

Bedspread-weight crochet cotton, one (225 yd ball) white
Size 7 steel crochet hook, or size required for gauge

Gauge:

Rnds 1 through 4 = 2 1/2" diameter

Instructions

Ch 4, join to form a ring.

Rnd 1: Ch 6 (counts as a dc and a ch-3 lp on this and following rnds), * dc in ring, ch 3; rep from * 6 times more; join in 3rd ch of beg ch-6: 8 ch-3 lps.

Rnd 2: Sl st in next ch-3 lp; ch 1, in same lp work (sc, 3 dc, sc): petal made; in each rem ch-3 lp work (sc, 3 dc, sc): petal made; join in first sc: 8 petals.

Rnd 3: Sl st in next 2 dc; ch 1, sc in same dc as last sl st made; (ch 6, sc in 2nd dc of next petal) 7 times; ch 6; join in first sc.

Rnd 4: Sl st in next ch-6 lp; ch 1, in same lp work (sc, hdc, 5 dc, hdc, sc): petal made; in each rem ch-6 lp work (sc, hdc, 5 dc, hdc, sc): petal made; join in first sc.

Rnd 5: Sl st in next hdc and in next 3 dc; ch 1, sc in same dc as last sl st made; (ch 9, sc in 3rd dc of next petal) 7 times; ch 9; join in first sc.

Rnd 6: Sl st in next ch-9 lp; ch 1, in same lp work (sc, hdc, 3 dc, 3 trc, 3 dc, hdc, sc): petal made; in each rem ch-9 lp work (sc, hdc, 3 dc, 3 trc, 3 dc, hdc, sc): petal made; join in first sc.

Rnd 7: Sl st in next hdc and in next dc; ch 7 (counts as a trc and a ch-3 lp), * sk next st, trc in next st, ch 3, sk next st, in next st work (trc, ch 3, trc): V-st made; (ch 3, sk next st, trc in next st) twice; sk next 4 sts, trc in next st, ch 3; rep from * 7 times more, ending last rep without working last trc and last ch 3; join in 4th ch of beg ch-7.

Rnd 8: Ch 1, sc in same ch as joining; (3 sc in next ch-3 lp, sc in next trc) 5 times; * sc in next trc, 3 sc in next ch-3 lp; sc in next trc, turn; ch 7, sl st in 9th sc; turn; over ch-7 lp work (sc, hdc, 7 dc, hdc, sc): scallop made; (3 sc in next ch-3 lp, sc in next trc) 4 times; rep from * 6 times more; join in first sc; sl st in next 4 sc; turn; ch 7, sl st in 9th st; turn; over ch-7 lp work (sc, hdc, 7 dc, hdc, sc): scallop made; sl st in next 6 sc: 8 scallops.

Rnd 9: Ch 15 (counts as a trc and a ch-11 lp), * sc in 4th dc of next scallop, ch 11, sk next 5 sc, trc in next sc, ch 11; rep from * 6 times more; sc in 4th dc of next scallop, ch 11; join in 4th ch of beg ch-15.

Rnd 10: Ch 7 (counts as a dc and ch-4 lp), * dc in next ch-11 lp, ch 4, in next sc work (3 dc, ch 1, dc, ch 1, 3 dc); ch 4, dc in next ch-11 lp, ch 4, dc in next trc, ch 4; rep from * 7 times more, ending last rep without working last dc and last ch 4; join in 3rd ch of beg ch-7.

Rnd 11: Sl st in next 2 chs of next ch-4 lp; ch 1, sc in same lp; * ch 5, sk next ch-4 lp, dc in next dc, 3 dc in next dc; dc in next dc, ch 2, in next dc work (dc, ch 1, dc); ch 2, dc in next dc, 3 dc in next dc; dc in next dc, ch 5, sk next ch-4 lp, sc in next ch-4 lp, ch 5, sc in next ch-4 lp; rep from * 7 times more, ending last rep without working last sc; join in first sc.

Rnd 12: Sl st in next 4 chs and in next dc; ch 3 (counts as a dc), dc in next 4 dc, * ch 2, in next ch-1 sp work (dc, ch 1, dc); ch 2, dc in next 5 dc, ch 3, sk next ch-4 lp, 7 dc in next ch-4 lp; ch 3, dc in next 5 dc; rep from * 7 times more, ending last rep without working last 5 dc; join in 3rd ch of beg ch-3.

Rnd 13: Ch 3, dc in next 2 dc, * ch 2, sk next 2 dc, 3 dc in next dc; ch 2, dc in next ch-1 sp, ch 2, 3 dc in next dc; ch 2, sk next 2 dc, dc in next 3 dc, ch 7, sk next 3 dc, sc in next dc, ch 7, sk next 3 dc, dc in next 3 dc; rep from * 7 times more, ending last rep without working last 3 dc; join in 3rd ch of beg ch-3.

Rnd 14: Sl st in next dc; ch 6, * sk next dc, dc in next dc, 3 dc in next dc; dc in next dc, ch 2, in next dc work (dc, ch 1, dc); ch 2, dc in next dc, 3 dc in next dc; dc in next dc, ch 3, sk next dc, dc in next dc, (ch 5, dc in next ch-7 lp) twice; ch 5, sk next dc, dc in next dc, ch 3; rep from * 7 times more, ending last rep without working last dc and last ch 3; join in 3rd ch of beg ch-6.

Rnd 15: Sl st in next 3 chs and in next dc; ch 3, dc in next 4 dc, * ch 2, sk next ch-2 sp, in next ch-1 sp work (dc, ch 1, dc); ch 2, sk next ch-2 sp, dc in next 5 dc, ch 5, sk next ch-3 lp, (dc in next ch-5 lp, ch 5) 3 times; dc in next 5 dc; rep from * 7 times more, ending last rep without working last 5 dc; join in 3rd ch of beg ch-3.

continued on page 59

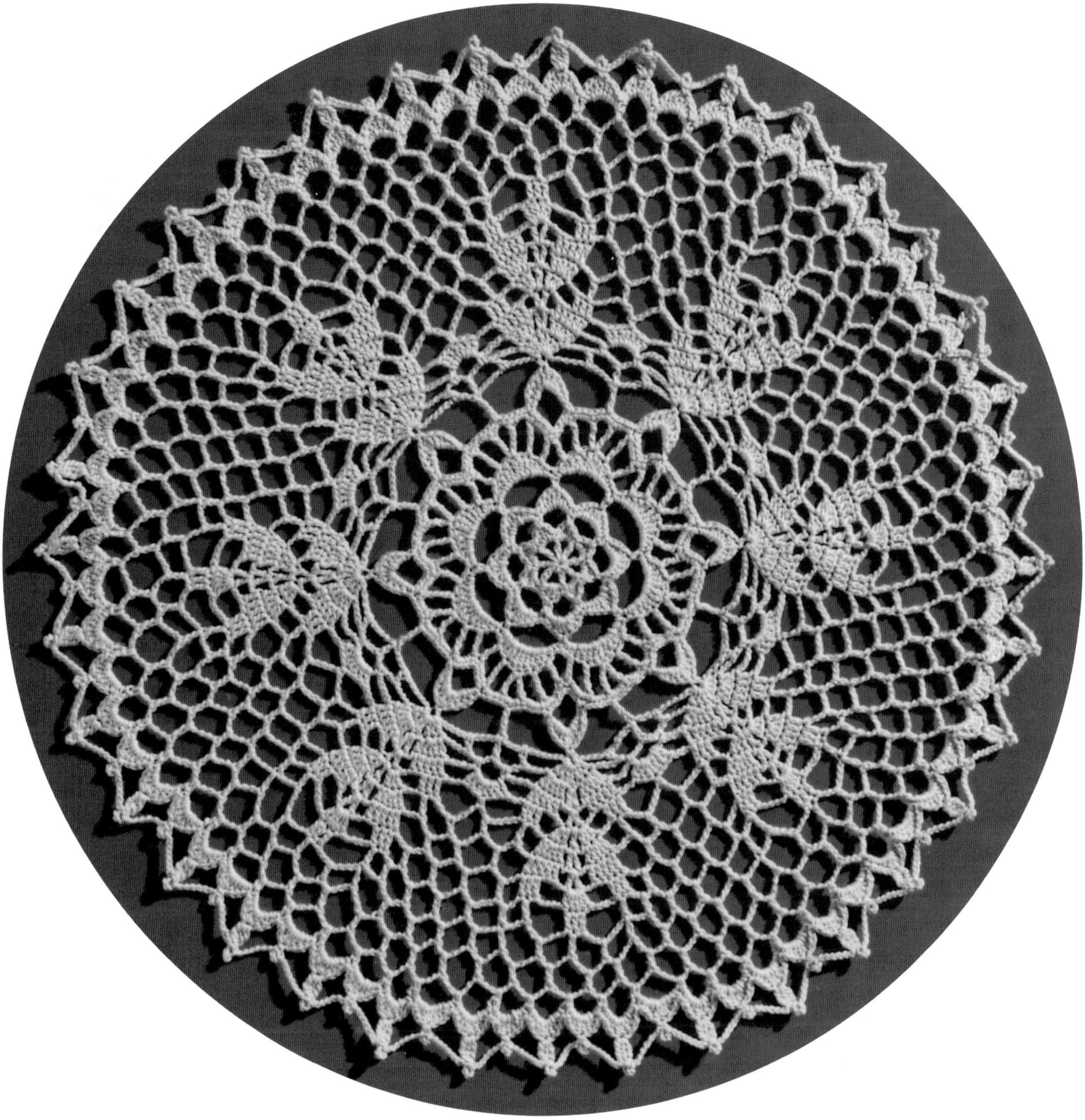

Rnd 16: Ch 3, dc in next 2 dc, * ch 5, sk next 2 dc and next ch-2 sp, 5 dc in next ch-1 sp; ch 5, sk next ch-2 sp and next 2 dc, dc in next 3 dc, ch 3, dc in next ch-5 lp, (ch 5, dc in next ch-5 lp) 3 times; ch 3, dc in next 3 dc; rep from * 7 times more, ending last rep without working last 3 dc; join in 3rd ch of beg ch-3.

Rnd 17: Sl st in next dc; ch 10 (counts as a dc and a ch-7 lp), * sk next dc, next ch-5 lp, and next dc; dc in next 3 dc, ch 7, sk next dc, next ch-5 lp, and next dc; dc in next dc, ch 3, dc in next ch-3 lp, (ch 5, dc in next ch-5 lp) 3 times; ch 5, dc in next ch-3 lp, ch 3, sk next dc, dc in next dc, ch 7; rep from * 7 times more, ending last rep without working last dc and last ch 7; join in 3rd ch of beg ch-10.

Rnd 18: Sl st in next 4 chs of next ch-7 lp and in same lp; ch 8 (counts as a dc and a ch-5 lp on this and following rnds), * sk next dc, dc in next dc, ch 5, dc in next ch-7 lp, ch 5, dc in next dc, (ch 5, dc in next ch-5 lp) 4 times; ch 5, dc in next dc, ch 5, dc in next ch-7 lp, ch 5; rep from * 7 times more, ending last rep without working last dc and last ch 5; join in 3rd ch of beg ch-8.

Rnd 19: Sl st in next 3 chs of next ch-5 lp and in same lp; ch 8, * dc in next ch-5 lp, ch 5: rep from * around; join in 3rd ch of beg ch-8.

Rnd 20: Sl st in next 3 chs of next ch-5 lp; ch 1, sc in same lp; ch 7, * sc in next ch-5 lp, ch 7; rep from * around; join in first sc.

Rnd 21: Sl st in next ch-7 lp; ch 1, in same lp work (sc, hdc, 5 dc, hdc, sc): scallop made; in each rem ch-7 lp work (sc, hdc, 5 dc, hdc, sc): scallop made; join in first sc.

Rnd 22: Sl st in next hdc and in next 3 dc; ch 4, keeping last lp of each trc on hook, 2 trc in same dc as last sl st made; YO and draw through all 3 lps on hook: beg cluster made; ch 4, sl st in top of beg cluster just made: picot made; * ch 7, sc in 3rd dc of next scallop; ch 4, sl st in sc just made: picot made; ch 7, keeping last lp of each trc on hook, 3 trc in 3rd dc of next scallop; YO and draw through all 4 lps on hook: cluster made; ch 4, sl st in top of cluster just made: picot made; rep from * around; join in top of beg cluster.

Finish off and weave in ends.

Metric Conversion Charts

INCHES INTO MILLIMETERS & CENTIMETERS *(Rounded off slightly)*

inches	mm	cm	inches	cm	inches	cm	inches	cm
1/8	3		5	12.5	21	53.5	38	96.5
1/4	6		5 1/2	14	22	56	39	99
3/8	10	1	6	15	23	58.5	40	101.5
1/2	13	1.3	7	18	24	61	41	104
5/8	15	1.5	8	20.5	25	63.5	42	106.5
3/4	20	2	9	23	26	66	43	109
7/8	22	2.2	10	25.5	27	68.5	44	112
1	25	2.5	11	28	28	71	45	114.5
1 1/4	32	3.2	12	30.5	29	73.5	46	117
1 1/2	38	3.8	13	33	30	76	47	119.5
1 3/4	45	4.5	14	35.5	31	79	48	122
2	50	5	15	38	32	81.5	49	124.5
2 1/2	65	6.5	16	40.5	33	84	50	127
3	75	7.5	17	43	34	86.5		
3 1/2	90	9	18	46	35	89		
4	100	10	19	48.5	36	91.5		
4 1/2	115	11.5	20	51	37	94		

mm – millimeter cm – centimeter

CROCHET HOOKS CONVERSION CHART

U.S.	1	2	3	4	5	6	7	8	9	10	11	12	13	14
English	3/0	2/0	1/0	1	1 1/2	2	2 1/2	3	4	5	5 1/2	6	6 1/2	7
Continental-mm	3	2.5		2		1.75	1.5	1.25	1	0.75		0.6		